The

GUIDE to the EMC DIRECTIVE

89/336/EEC

SECOND EDITION

Chris Marshman

The Institute of Electrical and Electronics Engineers, Inc.
New York

Reprinted with minor amendments 1996

First published by EPA Press, Wendens Ambo, UK. ISBN 0 9517362 7 2 2nd Edition

Exclusive North American distribution rights assigned to:

IEEE Press
445 Hoes Lane
Piscataway, NJ 08855-1331

ISBN 0-7803-1169-8
Order number: PC 5642

British Library Cataloguing in Publication Data

Marshman, Chris

Guide to the ElectromagneticCompatibility Directive 89/336/EEC, 2nd Edition
I. Title
621.31924

Printed in the United Kingdom by St Edmundsbury Press

to

Maryan, Anna, Jane and Harriet

Acknowledgements

The author gratefully acknowledges the following for their help in the preparation of this book.

Dr Katie Petty-Saphon, the publisher.

British Standards Institution for permission to reproduce figures from BS 800, BS 6667:Part2, IEC 801-4 and BS EN 55 011.

The Department of Trade and Industry, Manufacturing Technology division.

Electrical Review for permission to reproduce Figure 1.3.

The York Electronics Centre, University of York for access to data and for permission to reproduce the majority of the figures.

His wife for her encouragement, painstaking proof reading and converting in many instances 'gobble de gook' into something like English!

His wife and family for their overstretched patience during the preparation of this second edition.

Contents

Preface to the second edition

I am continually being asked to give advice regarding the European Community Directive on Electromagnetic Compatibility, 89/336/EEC, the implementing legislation and the implications for companies and advice on how they can comply with its requirements. This experience and expertise has been brought together in this book. The EMC Directive and the associated relevant standards, are necessarily written in a legalistic way, which makes it difficult for manufacturers of electrical and electronics products to interpret the implications. The situation is not helped because direct reference to a number of documents is required: the EMC Directive itself, the European Commission's explanatory document, the national legislation of the member states of the EC implementing the Directive and a number of harmonised European Standards (ENs).

This guide describes the key features of both the EMC Directive and the harmonised or relevant standards, it identifies the implications and, finally, provides guidance for achieving compliance by the use of an action plan and appropriate examples. By making the language less legalistic, it may also have become less precise, and so it is very important to appreciate that this book does not replace the EMC Directive and its associated documentation, but complements them. It is therefore recommended that the reader uses the comprehensive contents list as a starting point, as each subject area must be considered in context with reference to the implications. The index should be used as a secondary source as it provides an index of definitions which may be referred to should a chapter be read in isolation. The index also references the pages where a subject is covered in detail but due to the legally binding nature of the EMC Directive it is important to research each area thoroughly within its context.

In this second edition the material has been extensively revised and updated. Chapter 4 has been rewritten around the published UK regulations, a new chapter, Chapter 11, has been added describing an approach for those needing to establish compliance for large installed systems and much additional information has been included within the appendices. It should be noted that despite all the activity surrounding standards only one new standard has appeared in the OJ since August 1992, the generic emission strandard for the industrial environment, at the time of writing we are still awaiting the generic industrial immunity standard!

The text is aimed at the following readers:

- those involved in achieving compliance for electrical or electronic products: *the approvals engineer and quality managers.*

- those involved in marketing these products in the European Community: *marketing executives, importers and distributors of goods manufactured outside the EC.*

- those who will design or test equipment to meet specified levels of immunity or emission limits: *engineering or technical managers, electrical/electronics engineers and EMC test engineers.*

- those who have the task of being the 'signatory' on the declaration of conformity *(for small or medium size enterprises this may be the managing director).*

Fixes and EMC design are outside the scope of this book, although the message emphasised is that EMC must be considered as an integral part of the design process.

Every effort has been made to ensure that this book is correct in every detail, but it is possible that in such a fast changing regulatory and technical environment some errors may have crept in particularly due to the changes that are being introduced to standards by CISPR and CENELEC. Also there are a large number of standards and amendments which are currently being reviewed in committee before being finally adopted. The author would be grateful to receive comments and criticisms from readers, which should be sent to him through the publisher.

Chris Marshman
York
May, 1995

1

The European Community Directive on EMC — an Introduction

This chapter is provided to give a clear indication of the need for this guide and therefore the most efficient use of it. It provides a background to EMC and the requirement to control the unwanted electromagnetic emissions from electrical and electronic equipment and to confer to such equipment an intrinsic level of immunity to externally generated electromagnetic interference. Standards and regulations provide these controls. Hence there was a requirement for an EC Directive on EMC to enforce EMC controls for the 'single European market'. The new approach Directives and harmonisation of standards are explained along with the complex standards making machinery. This is set in the context of the European parliamentary and legislative process.

1.1 The Need for this Guide

In May 1989 the European Community (EC) published a Council Directive, 89/336/EEC[1], 'on the approximation of the laws of the Member States relating to electromagnetic compatibility (EMC)' to be effective from 1 January 1992. This document has significance for all manufacturers and installers of electrical and electronic equipment, either marketing products or providing services, within the European Union (EU) and the European Economic Area (EEA), as all electrical and electronic equipment falls within its scope. However the European Commission underestimated the task of implementing this Directive which is regarded as the most complex of the 'new approach' Directives, in consequence an amending Directive [92/31/EEC[2]] was published in April 1992 which allowed four years ,until January 1 1996, for the change from existing national regulations to the full implementation of the EMC Directive.

Due to the difficulties of interpreting the EMC Directive the European

Commission responded by publishing an 'explanatory document'[3]. As a 'new approach' Directive, the EMC Directive relies on the availability of harmonised standards. Some of these were available and the four year 'breathing space' has allowed the standards bodies to introduce new standards and to update others.

Manufacturers and distributors of electrical and electronic products in the EEA are faced with developing an understanding of what the EMC Directive means for them, how to comply with it and what will be the penalties if they fail to comply. After all the EMC Directive has to be legally implemented in all member states and non-compliance is a criminal offence!

This guide cannot replace the text of the EMC Directive, explanatory document, or the detailed national regulations of the member states. Also, it cannot replace the harmonised standards. However, it is the author's intention that this guide will help engineers and others within the electrical and electronics manufacturing, distribution and service industries, to understand and implement the requirements of the EMC Directive in the most cost effective way possible.

1.2 Background

The world's industrialised nations are becoming increasingly dependent upon the use of electrical and electronic equipment. In recent years the introduction of the microprocessor and microcomputer has brought about a technological revolution which has had far reaching effects in the home, in industry and in commerce. The proliferation of computing devices and electronics equipment has been accompanied by the declining use of electromechanical components. Since electronics components are inherently more susceptible to electromagnetic noise than their electro-mechanical counterparts, it becomes essential to determine the susceptibility of a system to electromagnetic interference (EMI) [Moy 1989[4]].

Typically, microcomputers work at clock frequencies of between 4 MHz and 25 MHz. Personal computers operating with a clock frequency of 100 MHz + are now commonplace. Their digital waveforms have fast rising and falling edges which may generate harmonics with a significant magnitude at frequencies greater than 1 GHz. Equipment incorporating a microcomputer, whilst performing its designed function, may also be emitting unwanted signals which may interfere with the operation of other electrical and electronic equipment. These unwanted signals are propagated as electromagnetic waves. Such signals are a potentially disruptive threat to electrical and electronic equipment, because they can create electromagnetic interference (EMI).

EMI is not exclusively produced by equipment incorporating a microcomputer. As the general use of electrical and electronic equipment increases, so does the number, and diversity, of EMI sources. The effects of EMI are of concern because they can cause key electrical and electronic systems to malfunction within, for example, manufacturing industry, communications or defence, and ultimately can present a direct threat to public health and safety if catastrophic failure occurs [Deb and Mukherjee 1985[5]].

Considering an electrical or electronic system, it will consist of a number of sub-systems between which there will be electromagnetic interactions. Inevitably the system will also interact with its environment. The electromagnetic interface between the system and its environment must be carefully controlled to avoid a significant mismatch.

This potential mismatch can be illustrated by considering the frequency spectrum, Figure 1.1: there are *bona fide* users of the spectrum - for example radio, TV and cellular radio networks. In addition there are the operating frequency ranges of, for example, audio and video signals and, at 50 to 60 Hz, power equipment. Switchmode power supplies can operate at frequencies from a few kilohertz up to around 200 kHz and may 'unintentionally' generate harmonics up to 10 MHz. Digital electronic equipment, including microcomputers, can operate from frequencies of tens of Hertz up to tens of MegaHertz with harmonics up to the GigaHertz region.

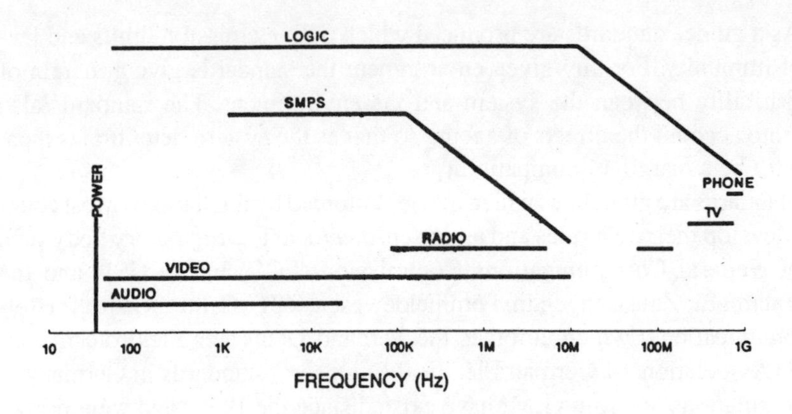

Figure 1.1 ***The frequency spectrum***

The harmonics created by switchmode power supplies and digital electronic equipment, (the unintentional users of the spectrum) are capable of interfering with *bona fide* spectrum users. As aspects of system design become increasingly susceptible to the effects of EMI, and as the electromagnetic environmental threat becomes more severe, it becomes necessary more than ever before to ensure compatiblity between systems.

Electromagnetic Compatibility (EMC) is the discipline which attempts to overcome or, at least, minimise the effects of mismatch between equipment and the operating environment in accordance with agreed specifications, standards and regulations.

Therefore in relation to EMI, the discipline of EMC has two aspects.

Firstly, it quantifies the extent to which any given electrical or electronic system can properly function without generating electromagnetic disturbances at a level that would cause a malfunction of other systems — this is the *EMISSION* aspect.

Secondly, it quantifies the extent to which any given electrical or electronic system can properly function without the risk of malfunction within a defined electromagnetic environment — this is the *IMMUNITY* aspect.

In practice, all EMI and EMC problems have elements which are unique to the particular system and environments in question. In order to achieve electromagnetic compatibility between electrical/electronic apparatus it is necessary to control:

i) emissions from equipment;
ii) the level of immunity of equipment to external electromagnetic disturbances.

As a guide, standards are produced which define emission limits and levels of immunity. For any given environment the standards give a margin of compatibility between the system and the environment. The standards also take into account the effects of ageing so that as the system deteriorates there will still be a margin of compatibility.

Standards are guidelines which may be enforced by regulations. Most countries develop their own rules and assign enforcement to a regulatory body such as the Federal Communications Commission (FCC) in the USA and the Zentralamt für Zulassungen im Fernmeldewesen, ZZF (Central Office for Telecommunications) which enforces the Verband Deutscher Elektrotechniker, VDE (Association of German Electrical Engineers) standards in Germany.

Regulations regarding EMI have existed since the 1950s and were primarily concerned with interference to radio and television receivers [Wireless Telegraphy Act 1949[6]]. Due to the increasing problems of interference in a wide range of commercial environments, more specific requirements and recommendations have been produced by national and international committees. Inevitably these have been devised from military specifications where, of necessity, the EMC problem has been tackled more vigorously; particularly in relation to susceptibility [Mil-Std-461[7] and Def Std 59/41[8]]. More recently however, the work of the International Electrotechnical Commission, IEC, largely through its CISPR committee, has had a much wider remit and seeks to set up standards which will be accepted worldwide. This will effectively remove obstacles to trade by ensuring that everyone is working to the same rules. Most standards follow the recommendations of CISPR the Comité International Spécial des Perturbations Radioélectriques (or International Special Committee on Radio Interference) for establishing emission limits, susceptibility levels and test procedures.

UK based companies have essentially only designed equipment with EMC in mind if their equipment's function is affected or if it is covered by a Statutory Instrument. Market forces have also dictated the necessity to meet FCC requirements when exporting equipment to the USA, and VDE standards when exporting to Germany. However from January 1996 equipment marketed within Europe, including the UK, is required to comply with the EC Directive on EMC [89/336/EEC[1]].

The European Community adopted this Directive in preparation for the single European Market. Its intention is to remove barriers to trade within the European Community, now the European Union (EU) and extended to include all states in the European Economic Area (EEA), by controlling emission and immunity levels. It will also provide an environment which will ensure the reliable operation of all electrical and electronic equipment. These controls will also apply to equipment imported into the EEA.

The regulations implementing this Directive have severe implications for a manufacturer whose equipment does not comply, both in terms of restricting the market for his product and legal action which might result in a substantial fine. The Directive differs from the regulations existing in some countries including the USA, by including the aspect of immunity as well as emission. Furthermore its scope is all-embracing: it is not simply aimed at a few specific areas of the electrical and electronics industry.

1.3 The Single European Market

The objective of creating a European 'common' market was an essential part of the 'Treaty of Rome' signed on 25 March 1957, which established the European Economic Community (EEC). In 1985 the European Community (EC), Heads of Government committed themselves in the 'Single European Act' (SEA), which came into operation on 1 July 1987, to complete the 'single market' progressively by 31 December 1992.

Despite the elimination of tariff and quota restrictions between member states, the common market envisaged in the Treaty of Rome has not been achieved. For example, the free movement of goods is impeded by technical barriers such as differing national product standards.

In 1988 the largest export market for the UK was the EC, accounting for approximately 50% of UK exports. This contrasts with 33% of UK exports in 1972. UK export of goods to the EC totalled £39 billion in 1987 [DTI 1989[9]]. The DTI suggested therefore that the UK would be presented with 'substantial new opportunities' on completion of the single market.

The SEA commited the EC to progressive establishment of the single market by 31 December 1992. The single market is defined as 'an area without internal frontiers in which free movement of goods, persons, services and capital is ensured in accordance with the provisions of the Treaty'. The SEA also implemented reforms to expedite the EC decision-making process by establishing majority voting to all the major areas of the single market programme.

The Community legislative process, which included the 'new cooperation procedure introduced by the SEA, is shown in Figure 1.2, reproduced from the DTI publication 'The single market - The Facts'[9] p11. There are four main community institutions: the Commission, the Council, the Parliament and the Court of Justice.

The ***Commission*** has three roles:

i) to propose Community policy and legislation;
ii) to execute decisions taken by the Council of Ministers and su pervise the running of community policies;
iii) to take action against Member States which do not comply with the Treaties.

Commissioners representing each EU member state are appointed for four years. Each Commissioner is in charge of an area of Community Policy. The Commission is divided into Directorates-General (DGs) and associated services. Each DG or service has a Commissioner responsible for its work. DGs of relevence are: DGIII — Internal Market and Industrial Affairs; DGXII — Science, Research and Development and DGXIII — Telecommunications, Information Industries and Innovation.

Each Commissioner formulates proposals within his or her area of responsibility aimed at implementing the Treaties. For example under the SEA five new Treaty Articles were introduced including Articles 100A and 100B —'Approximation of Laws' (Single Market). These proposals are then discussed by the Commissioners as a body who decide the final proposal and decisions are taken by majority vote.

The ***Council*** is the EC's decision-making body which adopts legislation on the basis of Commission proposals. The Council embraces the Council of Ministers, working groups of officials and the Committee of Permanent

Representatives of the Member States (COREPER). COREPER prepares the material discussed by the Council of Ministers.

The ***European Parliament*** is a directly elected body of approximately 500 members; with approximately 80 being from the UK. Its formal opinion is required on many proposals before they can be adopted by the Council.

The ***European Court of Justice*** **(ECJ)** rules on the interpretation and application of Community laws.The complement of judges includes one from each Member State. Judgements of the Court are binding in each Member State. There are four types of Community legislation:

Regulations,
Directives,
Decisions and
Recommendations/Opinions.

The latter have no binding force; Decisions are binding on those to whom they are addressed, *eg* Member States, companies or individuals; and Regulations are binding in all Member States taking precedence over National regulations. Directives are binding on Member States but specify a period within which they must be implemented. The method of implementation is at the discretion of the Member State. In the UK this may take the form of primary legislation or statutory instruments made under existing relevant specific powers. In itself a Directive does not have legal force in the Member States.

Under the Cooperation Procedure introduced by the SEA, the European Parliament gives its opinion on a Commission Proposal and on the 'common position' adopted by the Council of Ministers. In the latter case the Parliament has to give its opinion within 3 months by a majority of all MEPs. This allows the Parliament time to propose amendments after the Council has formed a view but before formal adoption as Community law, see Figure 1.2.

Most Member states have their own standards and laws controlling quality and safety of goods sold in the home market. These can be a barrier to trade when member states operate different standards or do not recognise each others arrangements for testing and certification. Lloyds Bank reported on a survey of UK firms carried out by the European Commission and found that of all the perceived barriers to trade, technical standards and regulations ranked highest [Lloyds Bank Economic Bulletin 1989[10]].

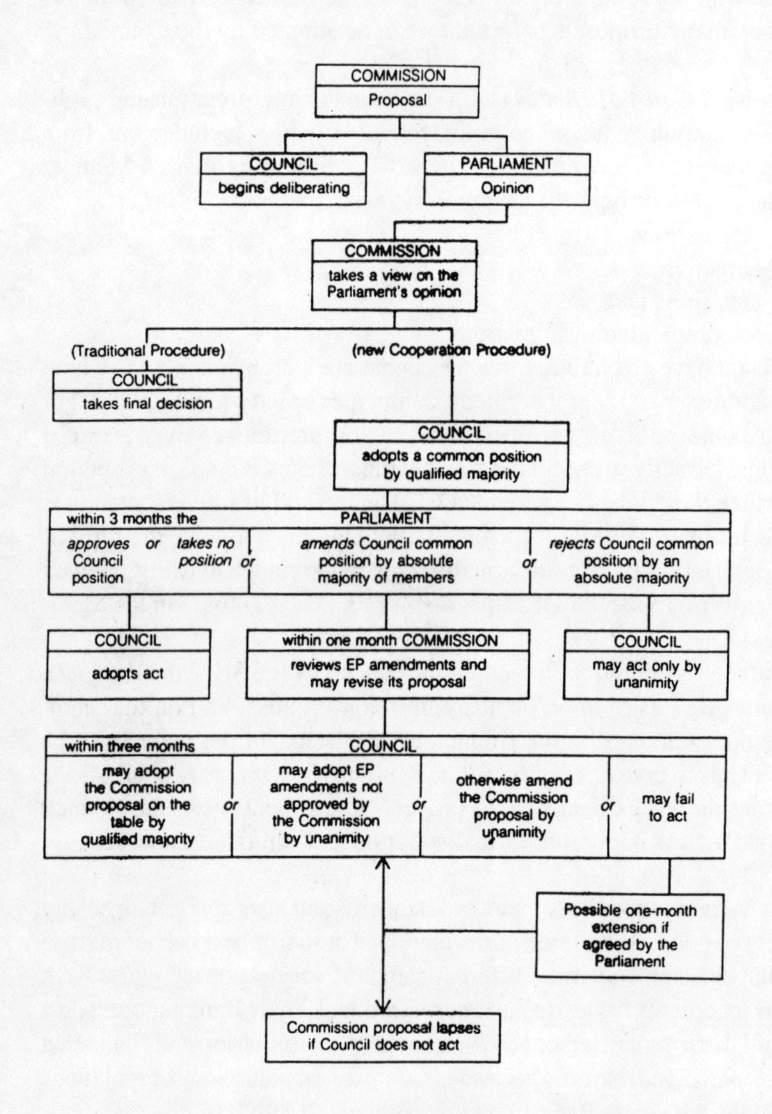

Figure 1.2 The Community legislative process
Reproduced with permission from the DTI

Since 1983 arrangements have been in force to prevent the creation of new technical barriers [Directive 83/189/EEC[11]]. Member States are now required to notify the Commission in advance of any proposals for technical regulations. In 1985 the Community agreed a *'New Approach to Technical Harmonisation'*. Previously EC Directives specified a mass of technical detail which required agreement. Under the new approach, Directives specify the essential requirements with the technical details being agreed separately under the auspices of the European Standards bodies, CEN (Comité Européen de Normalisation or European Standards Committee) and CENELEC (Comité Européen de Normalisation Electro-technique or European Standardisation Committee for Electrical Products).

These new approach Directives provide for total harmonisation. Products covered by them must meet their essential requirements and compliance with these will usually be demonstrated by meeting the relevant European standards (which replace individual national standards). Compliant products can then be marked with the 'CE' (Communauté Européenne) marking and be freely circulated within the EEA.

Areas covered by new approach Directives include: simple pressure vessels, toy safety, construction products, personal protective equipment, machinery safety, low voltage equipment, EMC, measuring instruments (*eg* non-automatic weighing instruments), medical devices, active implantable medical devices and gas appliances.

In addition to the harmonisation of standards, it is essential that there are common criteria for assessing the competence of national test and certification laboratories. Standards setting out these criteria have been established [EN 45 000 series[12]] and a Green Paper has been published setting out the operational framework for the European Organisation for Test and Certification (EOTC).

The countries of the European Free Trade Association (EFTA: Austria, Finland, Iceland, Norway, Sweden and Switzerland) together formed the EC's largest trading partner. The arrangements between the EC and EFTA were established by the 'Luxembourg Declaration' in April 1984. This framework called for action to improve the free circulation of industrial goods and highlighted the need to eliminate technical barriers. A full interchange of alterations to technical regulations was established and EFTA was fully represented on CENELEC committees from 1973 [Electrical Review 1991[13]]. These countries are now included within the EEA and several are already full members of the EU.

1.4 The Standards Machinery

Before the implications of the EMC Directive can be fully appreciated, an understanding of the standards making machinery is required. The standards bodies draft harmonised standards which are required as one of the means of demonstrating compliance with the EMC Directive's 'protection requirements'.

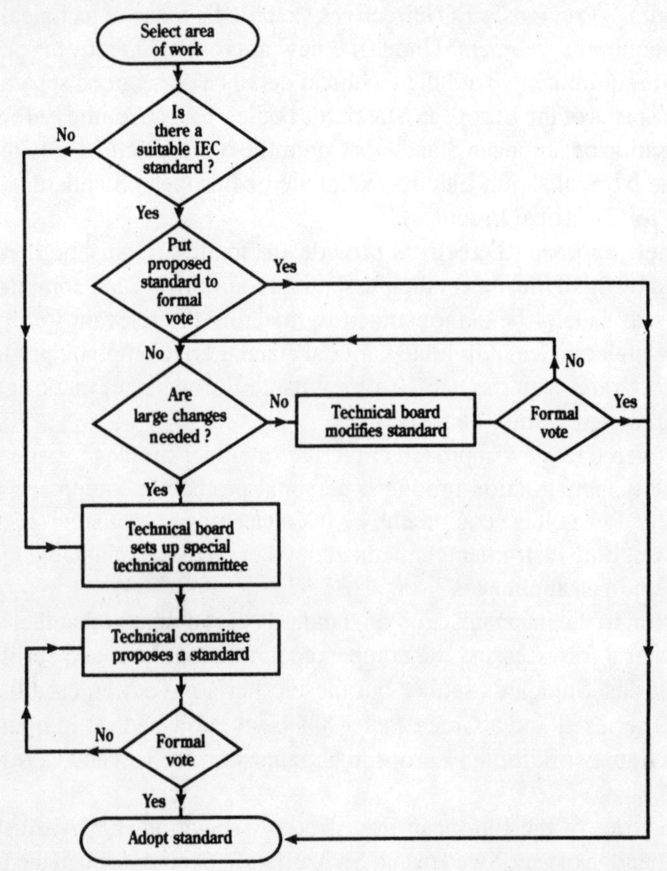

Figure 1.3 Steps involved in establishing a CENELEC standard
Reproduced with permission from *Electrical Review* 25th January 1991

CENELEC has the responsibility of producing harmonised European EMC standards. There are nine steps involved in establishing a CENELEC standard *(see* Figure 1.3 reproduced from Electrical Review 25 January 1991[13]):

1. ***An area of work is selected*** — CENELEC's technical board decides that a particular standard or group of standards is needed. The board comprises one representative from each national electrotechnical committee. Although most of CENELEC's work is related to EC Directives, only 10% is mandated by the EC [Electrical Review 1991]

2. ***CENELEC searches for a suitable international standard*** the most efficient way of producing a European standard is to adopt an IEC standard. Until adopted, an IEC standard only has the status of a recommendation. If a suitable IEC standard does not exist, a technical committee is set up. In the instance of an IEC standard being available (in 85% of cases) CENELEC proceeds with a formal vote [Electrical Review 1991].

3. ***The proposed standard is put out to formal vote*** — the national electrotechnical committees circulate the proposed standard for public comment, in the UK this process is initiated by an announcement in BSI News. After the period of public enquiry each national committee casts a vote weighted by the size of the country. Each national electrotechnical committee casts its vote and formulates a response to reflect the comments received. If the majority of votes are in favour the standard is adopted, otherwise CENELEC's technical board re-examines the proposed standard.

4. ***The technical board decides whether to set up a technical committee*** — the board sets up a technical committee if the proposed changes are major in nature, otherwise the board will modify the standard itself and puts it out for voting again.

5. ***The revised proposed standard is put out to vote again*** step 3. above is repeated.

6. ***CENELEC sets up a technical committee*** — the technical board appoints a chairman and a secretary and invites each national electrotechnical committee to send a delegation of up to five members. CENELEC has approximately 60 technical committees each with 30 to 40 members.

7. ***The technical committee drafts a proposed standard*** — this may be based on a variety of source materials for example a national standard, existing international work or information offered by a trade association. The technical committee may commission a smaller working group to produce a draft standard. Once a draft standard has been produced an iterative process is initiated, in which the technical committee delegates consult with their own national technical committee and then represent the national views back to the CENELEC technical committee, which will then modify the draft and the procedure is repeated. Once the proposed draft has a received a wide enough measure of support it will be put out for a formal vote.

8. ***The draft standard is put to formal vote*** — and the consultative process of step 3 carried out. If the standard is rejected it goes back to the technical committee for modification. Otherwise the standard is accepted.

9. ***The standard is adopted by CENELEC*** — it now has to be transposed as a national standard within a period defined by the technical board. This usually means that the standard will be published as a European standard or Euro Norm (EN) and each national standards body will publish a 'harmonised' equivalent. In the case of BSI these are now being issued as a BS EN for example BS EN 55 011.

The CENELEC EMC technical committee is TC110 which has responsibility to produce harmonised standards through which compliance with the EMC Directive can be demonstrated. This includes 'Generic' and 'Product or Product Family Specific' standards for aspects of emissions and immunity.

It is clear from the outline of the CENELEC standards making procedure that there is a dependence on IEC standards. The aim of the IEC is 'to promote international co-operation on all questions of standardisation and related matters in the fields of electrical and electronic engineering and thus promote international understanding' [Kay 1990[14]]. The IEC has 41 member countries and closely co-operates with the International Organisation for Standardisation (ISO). Of the IEC's 199 Technical Committees (TC's) and Sub-committees (SC's), 50 are involved with EMC aspects [Kay 1990[14]]. Two IEC committees are devoted to EMC work TC 77 (EMC between electrical equipment including networks) and CISPR.

At present approximately 90% of CENELEC EMC standards are in line with IEC publications with and without modifications [Kay 1990[14]]. Because of the pressures on CENELEC to produce standards which enable products to be declared compliant with the relevant new approach Directives, the IEC and CENELEC have agreed to increase co-operation to speed up the standards making process [IEC Bulletin 1990[15]]. The crucial area is in 'parallel voting', where draft international standards can be voted on by the IEC and CENELEC, with the aim of achieving international and regional standardisation simultaneously. CENELEC will also make its standards requirements known to the IEC technical committees.

The IEC co-ordinates its EMC work through the Advisory Committee on EMC (ACEC). The ACEC has established the principle of defining 'basic' publications (the IEC 1000 series of standards) which will ultimately be called up by specific publications [Kay 1990[14]]. This principle is also being used by CENELEC for its generic standards and can be expected to be used for all future EMC standards.

1.5 The Guide

1.5.1 The objectives of this guide

This guide is concerned with the implications of the EMC Directive on the electrical and electronics industry. These implications are of a technical, legal and economic nature. It is also concerned with interpreting the EMC Directive and associated documentation and providing guidance on compliance.

The technical implications of the EMC Directive occur in two distinct areas. The first is in defining whether a particular product is required to comply and choosing the route to demonstrate compliance. The second area is concerned with the technical implications of the harmonised standards. Some of the implications for the second area are also economic in nature, *eg* investment in test facilities. The implications for industry in both these areas are discussed.

The EMC Directive will become legally binding in all member states of the EC. The responsibilities of manufacturers and the administrations in each member state are identified. The proposed policing of the EMC Regulations in the UK and possible penalties are discussed.

The demand for EMC services, consultancy and test equipment is considered using, as a benchmark, the 1989 report by W S Atkins[16] for the DTI and the report published by John Moore Associates[17] in 1993.

The EMC Directive is widely regarded as the most complex of the new approach Directives, therefore as well as attempting to define the implications for industry, a step-by-step plan of action has been identified enabling manufacturers to demonstrate compliance for their products. The plan that has been developed (Chapter 10) draws heavily on the author's experiences of managing the the University of York's commercial EMC activities since 1989, on the work carried out for a variety of clients and products, and from the abundant questions received when lecturing on the subject of the EMC Directive at short courses, EMC awareness seminars and conferences.

1.5.2 Using this guide

A complete study of this book should prove to be of great value to those who require an understanding of the EMC Directive's implications. However, it is appreciated that the majority of readers will want to use the book for reference purposes. Therefore a comprehensive index and contents' list have been provided, along with a summary of individual standards which are appended. The 'references' section at the end of each chapter provides details of the original reference sources which may be used for further reading.

This book is not intended as an EMC textbook and does not cover electromagnetic fundamentals or EMC design principles. For readers requiring such a information they are referred to the books by Clayton R Paul, 'Introduction to Electromagnetic Compatibility', John Wiley & Sons, Inc., and David A Weston, 'Elecromagnetic Compatibility Principles and Applications', Marcel Dekker, Inc., and also to the Institution of Electrical Engineers (IEE) video based Distance Learning course on EMC.

References

1. 89/336/EEC Council Directive 'On the approximation of laws of Member States relating to electromagnetic compatibility', Official Journal of the European Communities No.139 25 May 1989, pp 19-26
2. 92/31/EEC Council Directive of 28 April 1992 'amending Directive 89/336/EEC on the approximation of laws of the Member States relating to electromagnetic compatibility', Official Journal of the European Communities No L 126/11, 12 May 1992
3. EC explanatory document on Council Directive 89/336/EEC, 111/4060/91/EN-Rev. 1, 1991
4. K P Moy 'EMC related issues for power electronics', IEEE TH0299-8/89/0000-0046, 1989, pp 46-53
5. G K Deb, M Mukherjee 'EM susceptibility studies and measurements on electro-explosive devices', IEEE 1985
6. Wireless Telegraphy Acts 1949 and 1967.
7. Mil-Std-461 'Military Standard Electromagnetic Emission and Susceptibility Requirements for the Control of Electromagnetic Interference', Department of Defense, USA
8 Def Std 59/41 'Defence Standard 59-41 Electromagnetic Compatibility', Ministry of Defence, UK
9. Department of Trade and Industry and the Central Office of Information 'The Single Market - The Facts', HMSO Dd 8940414 INDY J0631 RP, November 1989
10. Lloyds Bank '1992 winners and losers', Economic Bulletin Number 121, ISSN 0261-0175, January 1989
11. 83/189/EEC Council Directive 1983
12. EN 45 000 Series of standards, BS equivalent BS 7500 series, 'Criteria for the operation of testing laboratories', British Standard publications, 1989
13. D Martin 'Deciphering CENELEC' Electrical Review 25 January - 7 February 1991, pp16-17
14. R Kay 'Co-ordination of IEC standards on EMC and the importance of participating in standards work', IEE Seventh International Conference on EMC, Conference Publication 326, 1990
15. International Electrotechnical Commission 'IEC and CENELEC - The next stage of cooperation', IEC Bulletin Vol. XXIII No.122 - March/April 1990
16. W S Atkins 'The UK Market for EMC Testing and Consultancy Services', 1989
17. John Moore Associates Limited 'Electromagnetic compatibility testing in the UK', ISBN 0 9521797 0 9, June 1993

2

The European Community Directive on EMC

The EMC Directive is widely regarded as the most complex of the new approach Directives. It is therefore essential that its implications are fully understood by all manufacturers of electrical and electronics equipment and advisers to them. This means senior management in technical, financial and marketing areas. This chapter unravels the 'legalese' and provides an understanding of the EMC Directive by explaining the essential protection requirements and the scope of the equipment and phenomena covered. The routes to compliance are described, Self-Certification and the Technical Construction File. Also product certification and marking requirements are described. The responsibilities of EC Member States are defined.

2.1 General

From 1st January 1992 all electrical and electronic equipment 'placed on the market and taken into service' was required to comply with the objectives of the European Community EMC Directive[1] agreed in May 1989. This applies to both new and existing designs *ie*: designs currently under development and existing designs which are still in production and being marketed. The EMC Directive included a one year transitional period to allow for the introduction of standards. Following the introduction of the EMC Directive it was, in practice, found that this period was too short and an 'amending' Directive [92/31/EEC[2]] was drafted by the European Commission proposing the extension of the transitional period to four years. This amending Directive was adopted and was published in the Official Journal of the European Communities (OJ) on the 28 April 1992.

The EMC Directive was seen as an essential precursor to the establishment of the *Single European Market* and is intended to provide an environment for the reliable operation of all electrical/electronic equipment.

The EMC Directive is a 'new approach' Directive. The objectives defined by a new approach Directive are mandatory, whilst standards are not themselves binding and are only defined as a means of demonstrating that compliance with the objectives has been achieved. The standards can therefore be adapted to take account of technological progress, thus ensuring that development is not stifled.

2.2 The Protection Requirements of the EMC Directive [Article 4]

The objectives of the EMC Directive are encompassed by the protection requirements placed on electrical and electronic apparatus which must be constructed to ensure:

a) 'the electromagnetic disturbance it generates does not exceed a level allowing radio and telecommunications equipment and other apparatus to operate as intended;'

b) 'the apparatus has an adequate level of intrinsic immunity of electromagnetic disturbance to enable it to operate as intended.'

2.3 Scope of the Directive [Articles 1 & 2]

All electrical and electronic apparatus together with equipment and installations containing electrical/electronic components, are without exception deemed to be within the scope of the EMC Directive. The existing Directives[3] defining the EMC provisions for domestic equipment and luminaires will be absorbed into the EMC Directive. During 1991 the European Commission drafted an 'explanatory document', DG111/4060/91[4], which clarifies the scope and this is broadly in line with the Department of Trade and Industry's (DTI) consultative document[5], which was circulated in November 1990. A listing is made of all equipment which is covered but which should, 'not be regarded as restrictive', this is based on Annex III to the EMC Directive. The interpretation of components and installations are also included in the explanatory document[4] (*see* Chapter 3).

Just as the scope of the EMC Directive is wide ranging, the definitions of electromagnetic disturbances are all embracing and cover:

- conducted and radiated emissions,
- conducted and radiated immunity,
- mains disturbances,
- electrostatic discharge (ESD) and
- lightning induced surges.

This inclusion of all electromagnetic phenomena has been described as covering the frequency range of 'dc to daylight'!

The EMC Directive excludes equipment covered by other Directives with EMC provisions. This includes vehicle spark ignition systems[6] and non-automatic weighing instruments[7]. It should be noted that where separate provisions exist, but cover only certain aspects of EM disturbances (*eg* immunity to radiated interference), equipment is still required to comply with the EMC Directive in respect of the other aspects (*eg* radiated emissions). Also excluded is amateur radio equipment which is not commercially available.

2.4 Compliance with the Protection Requirements [Article 10]

Manufacturers and also distributors of imports from outside the European Economic Area are required to provide a declaration that their equipment complies with the protection requirements of the Directive. However, for demonstrating compliance three routes are described.

2.4.1 Standards or self-certification route [Article 10 (1)]

This is likely to become the primary route by which manufacturers (or their authorised representatives) demonstrate compliance with the EMC Directive. It is also the simplest method and is achieved by satisfying 'relevant standards' which is likely to be demonstrated either through in-house testing, or contracting the tests to an independent test house. The EMC Directive delegates responsibility for standards to CENELEC, which is required to produce standards in the form of European Standards or Euro Norms (EN). These generally follow the recommendations of CISPR and are defined as 'relevant standards'. Each national standards body is required to produce standards harmonised with the appropriate Euro Norm.

To be more precise a relevant standard is defined by the EMC Directive [Article 7] as a national standard, which is harmonised with a standard whose reference number has been published in the Official Journal of the European Communities (OJ).

In the absence of a European Standard, compliance with an existing national standard will suffice if the particular standard is accepted by the Commission and published in the Official Journal. However this is only likely to be an interim measure.

During the amended transitional period from 1 January 1992 to 31 December 1995, manufacturers may choose to comply with the EMC Directive *or* with existing national regulations. Manufacturers who can demonstrate compliance for their products may take full advantage of the single market by being able freely to market their products throughout the EEA. If products do not comply or a manufacturer chooses not to seek compliance, then these may continue to be marketed subject to the existing national regulations *eg*: the ZZF enforced VDE standards in Germany (*see* Chapter 12). This choice may well depend upon the availability of harmonised standards or national standards which have been adopted by the Commission as well as commercial considerations. It should be noted that the manufacturer is required to comply with the *protection requirements* of the EMC Directive not with particular standards. The EMC Directive specifically refers to the possible shortcomings of standards [Article 9].

2.4.2 Technical file [Article 10 (2)]

The alternative method is to produce and hold a 'technical file', to be available for inspection by the 'competent national authority' responsible for policing the EMC Directive. This form of certification implies that the technical file should demonstrate conformity with the *protection requirements* of the EMC Directive.

The technical file or technical construction file (TCF) should contain a description of the equipment and the EMC provisions made, it must also include a technical report or certificate from a 'competent body'. The technical report may be based on a theoretical study and/or appropriate tests. The manufacturer (or agent) is required to hold the TCF at the disposal of the enforcement authorities of any member state for a period of ten years. This route for claiming compliance will be obligatory after 1 January 1996 if there is no appropriate relevant standard.

Annex II to the EMC Directive lists the requirements of a competent body. In the UK, the Department of Trade and Industry requires NAMAS accreditation for laboratory facilities as one of the qualifications for competent body status. The DTI has published a list of competent bodies [DTI[11]].

NAMAS, the UK National Measurement Accreditation Service, is a division of the National Physical Laboratory (NPL), whose purpose is to assess and accredit laboratories which have demonstrated their competence to perform defined measurements within prescribed limits of uncertainty. NAMAS was formed by an amalgamation of the British Calibration Service (BCS) and the National Testing Laboratory Accreditation Scheme (NATLAS). A manufacturer may obtain accreditation for his own test facilities if these satisfy NAMAS requirements. In this instance it would be necessary to demonstrate that the testing facility is not compromised by pressures from the production side of the manufacturer's operations. For example the management 'tree' should indicate the independence of the testing facility and show no direct line of command from the control of production.

The UK interpretation of the Technical Construction File (TCF) route to compliance suggests that a manufacturer can prepare his own TCF and then submit this to a competent body. The competent body may accept this file and issue a certificate, or it may insist on further evidence of compliance, for example further testing and then issuing a technical report. The DTI suggest that a competent body should be consulted prior to preparation of a TCF.

In the absence of appropriate European or national standards, it may be necessary for a manufacturer or the competent body to consult with the national authority in order to carry out testing to standards which are not specifically for the application, but which the authority will accept in a technical file.

The DTI have prepared guidelines on the preparation of a TCF[12], the suggested contents is as follows:

Part I: Description of the apparatus:
- i) Identification of apparatus;
- ii) A technical description;

Part II: Procedures used to ensure conformity of the apparatus to the Protection requirements:
- i) A technical rationale;
- ii) Details of significant design elements;
- iii) Test evidence where appropriate.

Part III: A report or certificate from a 'Competent Body'.

2.4.3 EC type-examination certificate [Article 10 (4) and (5)]

The EMC Directive originally covered equipment used to send, process or receive information, *ie* Telecommunications Terminal Equipment (TTE), that is, equipment directly or indirectly connected to a public telecommunications network. It was a requirement for this equipment to be assessed by a 'notified body' which would issue an EC Type-examination certificate [Article 10 (4)]. This is no longer the case as a Directive specifically covering TTE was adopted in 1991, 91/263/EEC[8]. However, an EC Type-examination certificate is required for radiocommunications transmitters excepting those used by radio amateurs which are not commercially available [Article 10 (5)]. A notified body will, to all intents and purposes, be the same as a competent body but will be appointed by the national authority and 'notified' to the European Commission.

2.5 EC Declaration of Conformity

A manufacturer or import agent must hold an 'EC declaration of conformity' for equipment to be placed on the market. This declaration must contain: a description of the apparatus to which it refers, the specifications under which conformity is declared, identification of the signatory empowered to bind the manufacturer or agent and, where appropriate, reference to the 'EC type-examination certificate' issued by the notified body. The declaration must be kept available to the enforcement authorities for inspection purposes for a period of ten years after the last product has been manufactured.

The manufacturer or agent established within the Community must apply the Communauté Européenne (CE) marking to the equipment or else to its operating instructions, the guarantee certificate or its packaging. The CE marking should be affixed visibly, legibly and indelibly. It is also prohibited to affix marks which can be confused with the CE marking. Use of the CE marking indicates that the equipment complies with all the New Approach Directives applying to it and legally enforced, for example electronic toys must comply with both the toy and EMC Directives on 1 January 1996.

Conformity assessment procedures are laid down in the Council Decision 90/683/EEC, OJ L 380/13, 31 December 1990 (*see* Chapter 3).

A suggested proforma for a declaration of conformity is given at the end of the chapter.

2.6 Responsibilities of the Member States

Apparatus complying with the protection requirements of the Directive, that is to say bearing the CE marking, must not be impeded from being placed on the market [Article 5]. However, if the responsible administration of a Member State finds that apparatus does not comply [Article 9.1], then the apparatus must be *withdrawn from the market, its placing on the market prohibited or its free movement restricted.* The European Commission is then immediately informed.

The Commission is required to consult with the parties concerned as soon as possible. After the consultative process is completed and the Commission finds the action is justified, it must inform the Member State initiating the action and all the other Member State administrations. *This effectively will 'ban' the equipment throughout the European Community.* For UK manufacturers and distributors the implication is that this ban will include the UK even though the complaint originated in another member state.

Where non-compliant equipment is being marketed under a declaration of conformity the Member State responsible for identifying non-compliance is required to take *appropriate action against* the 'author of the attestation' in other words *the signatory on the declaration of conformity*. The Member State also has the responsibility of informing the Commission and the other Member States of its actions. It is then the responsibility of the Commission to ensure that the Member States are kept informed of the progress and outcome of any action taken [Article 9 (4)].

A Member State identifying non-compliant equipment must indicate to the Commission its reasons for believing it to be non-compliant. Three reasons for non-compliance are identified by the EMC Directive:

a) failure to meet the requirements of the harmonised standards
b) incorrect application of the harmonised standards
c) shortcomings in the standards themselves

Where the failure to comply is identified as being due to the shortcomings of standards, the Commission is required to bring the matter before the 'standing committee' (set up by Directive 83/189/EEC[9]) within two months, if the Member State taking the action intends to proceed. The Commission will consider the standard(s) in question and carry out appropriate consultation. The standing committee will then 'deliver an opinion without delay'. Dependent upon this opinion the Commission will inform the Member States whether or not it is necessary to withdraw the standard(s) in question. It should be noted that the *manufacturer is not excused because of the shortcomings of a standard*; after all he is required to comply with the protection requirements not the standards!

2.7 UK Implementation

In the UK the Wireless Telegraphy Acts[10] (WT Acts) exist to preserve the quality of radio communications. These regulations were updated to include provisions for CB radios, portable tools and fluorescent lighting; taking account of earlier European Directives. This legislation is *responsive* whereas the EMC Directive, which controls the conditions for apparatus to be placed on the market is *preventative*. Hence changes to the legislation were required. The new legislation had to include immunity requirements which had not been implemented by the earlier Acts. The DTI considered that new primary legislation was required because the task of modifying the existing Acts was too great but the new bill, originally scheduled to be passed through Parliament during the summer of 1991, was delayed pending the EC's decision on the amending Directive. The new legislation was passed by Parliament on 7 October 1992 and became effective on 28 October 1992. The UK EMC regulations are described in Chapter 4.

Enforcement of the WT Acts is currently carried out by the DTI's Radio Investigation Service, which is complaint driven. It is the DTI's intention that the new legislation, implementing the EMC Directive, will also be complaint driven. The enforcement authorities will be the weights and measures authorities (for other than radiocommunications equipment). This would appear to be inadequate when the nature of the Directive is considered. Random checks on equipment as a deterrent will actually be necessary if the spirit if the Directive is to be maintained. The UK implementation is discussed in detail in Chapter 4.

2.8 Summary

The EC Directive on EMC was to become binding on 1 January 1992. It did not however become legally binding in all Member States until late 1992. It applies to all electrical/electronic equipment and to all electromagnetic phenomena. It is a new-approach Directive defining the essential protection requirements with which compliance must be demonstrated. Three methods are described for demonstrating compliance. The first is by self-certifying that equipment conforms to harmonised standards, the second by compiling a technical construction file containing either a certificate or technical report from a competent body and the third by obtaining an EC type-examination certificate (radiocommunications transmitters). Compliant equipment may then be marked with the CE marking and freely marketed throughout the EEA. Each member state is required to implement the EMC Directive with appropriate legislation and establish means to enforce it.

The explanatory document which has been issued by the Commission is considered in the next chapter. It clarifies some of interpretational difficulties of the EMC Directive. In subsequent chapters the implications of the EMC Directive and the harmonised standards are discussed in detail.

References

1. Council Directive of 3 May 1989 on the approximation of the laws of the Member States relating to electromagnetic compatibility (EMC), 89/336/EEC, OJ L139 of 23.05.89, pp 19-26
2. 92/31/EEC Council Directive of 28 April 1992 'amending Directive 89/336/EEC on the approximation of laws of the Member States relating to electromagnetic compatibility', Official Journal of the European Communities No L 126/11, 12 May 1992
3. Directive '- relating to radio interference caused by household appliances, *etc*', 76/889/EEC, Directive '- relating to the suppression of radio interference with regard to fluorescent lighting luminaires', 76/890/EEC
4. EC explanatory document on Council Directive 89/336/EEC, 111/4060/91/EN-Rev. 1
5. 'Electrical Interference: a Consultative Document', DTI/PUB207/10k 10.89

6. Spark Ignition Systems - Motor Vehicles Directive 72/245/EEC and Agricultural and Forestry Tractors, Directive 75/322/EEC
7. Non-automatic weighing instruments, the immunity aspect is covered by Annex 1, para. 8(2), of Directive 90/384/EEC
8. Telecommunications Terminal Equipment Directive, 91/263/EEC, OJ L 128, 23 May 1991
9. Council Directive 83/189/EEC, 1983
10. Wireless Telegraphy Acts 1949 and 1967
11. DTI 'List of Competent Bodies appointed by the Secretary of State for Trade and Industry for conformity with the EC Directive 89/336/EEC through the Technical Construction File route', October 1992
12. DTI 'Guidance Document on the Preparation of a Technical Construction File as required by EC Directive 89/336', October 1992

Declaration of Conformity
(proforma)

Equipment name/type/number ____________________

Manufacturer. ____________________

Address ____________________

____

European Agent ____________________

European Standard(s)/TCF/Type Examination Certificate ____________________

Level/class ____________________

Conformity Criteria ____________________

Description of Equipment ____________________

I certify that the apparatus identified above conforms to the requirements of

Council Directive 89/336/EEC ____________________

Signed ____________________

Date ____________________

Position/status ____________________

Company name ____________________

3

The Explanatory Document Explained

In 1991 the European Commission published an explanatory document in an attempt to interpret the EMC Directive. The key points made in this document are expressed clearly in this chapter in order to enable technical management and approvals engineers to clarify what is meant in the EMC Directive by its scope, the terms 'placed on the market' and 'taken into service'. The explanatory document also goes some way to answer questions which have been raised in respect of one-off customised equipment, imports, second-hand equipment, systems, installations and components: these aspects are all described. A table is also provided identifying the EMC provisions within other Directives. The explanatory document and hence this chapter also clarifies the conformity procedures. Therefore this chapter improves the reader's understanding of the Commission's intentions behind the EMC Directive and clarifies a number of issues affecting product compliance.

3.1 Introduction

The EC Directive on EMC [89/336/EEC[1]] was published in May 1989 and originally required member states to transpose it into national law by 1 July 1991. In November 1989 the Department for Trade and Industry (DTI) circulated a Consultative Document[2] which outlined its intentions for the UK legislation and also asked for the industry's response. Companies and trade associations did respond and as a result of this exercise the DTI pressed the European Commission for more time for the introduction of the EMC Directive (to allow for the preparation of harmonised standards) and for guidance on various aspects of its interpretation. At a meeting in Brussels on 4 December 1990 this resulted in the European Commission agreeing to draft an amending Directive extending the transitional period to 31 December 1995. There was also agreement to provide guidance on the interpretation of the EMC Directive, which resulted in the 'Explanatory Document'[3] which became available in November 1991. This actually includes much of the DTI's interpretation explained in the consultative document[2].

The amending Directive [92/31/EEC[4]] was adopted on 28 April 1992 and published in the Official Journal of the European Communities (OJ) No L 126/11 on 12 May 1992. This means that the transitional period is now extended to 31 December 1995. During this period manufacturers may choose to comply with the EMC Directive, apply the CE marking and take full advantage of the single market even if existing national regulations still in force are more stringent. Alternatively a manufacturer may choose to comply with the national regulations in force within a member state from 30 June 1992.

The EMC Directive is now widely regarded as the most complex of the 'new approach' Directives, because of the breadth of its scope and the variety of products covered. It is therefore the aim of the explanatory document to clarify certain aspects and procedures to ensure the uniform application of the Directive. This should remove 'obstacles and difficulties' experienced by manufacturers, agents and any other 'concerned groups'. Unfortunately the language of the explanatory document is not helpful to understanding and therefore this chapter will attempt to clarify the key issues.

3.2 Definitions

3.2.1 Placed on the market

Placed on the market' referred to in article 3 of the EMC Directive is defined to be: the *first time* a product is made available for distribution and/or use in the European Economic Area (EEA) market. This product may be either manufactured within the EEA or imported from outside the EEA and the product is supplied either for payment or free of charge.

The emphasis is on the *first time* the product passes from the manufacturing stage or the product importation stage, to the distribution network or to the user. The EMC Directive applies only to new products manufactured *within* the EEA, but it applies to both new and 'secondhand' equipment imported from *outside* the Community.

The explanatory document provides further explanation of the 'availability of the product':

- it considers in the first instance the 'disposal' of the product. Disposal means the transfer of ownership, the physical transfer from the manufacturer, or from the manufacturer's agent ('authorised representative' established within the EEA), or the importer to the *distributor*. At the time of disposal the product must comply with the requirements of the EMC Directive. It is therefore the responsibility of the manufacturer or his agent or the importer to ensure that the product complies with the EMC Directive. The transfer will be any type of commercial transaction including: sale, loan, hire, leasing, or gift.

- the manufacturer, his agent, or the importer may also offer a product within their own distribution chain for *direct disposal to the user*. In this case the equipment must comply when it passes to the user. If a product is offered in a catalogue it is only when it is actually made available that it is placed on the market. However this implies that it was intended that the product would meet the provisions of the EMC Directive when it was offered.

- an organisation manufacturing or importing a product for its *own use* must ensure that it complies with the EMC Directive. This is because the EMC Directive will apply when the equipment is *first used*. It is considered that this example of 'placing on the market' should be included under 'taking into service' only, as the product is not 'seen' by the market at all.

Placed on the market' is exclusively concerned with the 'disposal' of products either into the distribution chain or direct to the user within the EEA, therefore the following are excluded from complying with the EMC Directive:

- importation into the EEA for the purpose of re-exportation.

- manufacture of a product within the EEA for export to a country outside the EEA.

- the display of a product at a trade fair or exhibition. However if the product is subsequently 'placed on the market' it will be required to comply when it enters the distribution chain or is passed to a user, as previously discussed.

- strictly the provisions of the EMC Directive do not apply to products which are passed from a manufacturer to his authorised representative or importer as *they* are responsible for demonstrating compliance of the product with the EMC Directive.

The explanatory document also states that *'Placed on the market' covers every individual, physically existing finished product within the scope of the EMC Directive, regardless of the time and place of manufacture and whether it was individually made or mass-produced.*

This means that if a manufacturer holding a stock of products predating the Directive's implementation date passes them into the distribution chain during the post-implementation period, then at the date of disposal this becomes the first time the product is available and therefore it is required to comply. Conversely a stock of pre-implementation manufactured products does not need to comply if it is held by a distributor who disposes of them to users in the post-implementation period, because the products *entered the distribution chain prior* to the implementation date.

The reference to 'every, individual....product' does not imply that every item produced has to be tested. But it is also clear that compliance has to be claimed for every product.

There is no distinction within the EMC Directive, between mass produced items and custom built 'one-off' products. The DTI's view, expressed in the consultative document, was that the custom one-off may not in fact be placed on the market and is therefore covered under the provisions of 'taken into service'.

3.2.2 Taken into service

'Taken into service' is referred to in article 3 of the EMC Directive. It is defined by the explanatory document as meaning: *the first use in the Community (EEA) of a product by its final user.*

The explanatory document states that 'where a user manufactures or imports a product covered by the EMC Directive for his *own use* the product must meet the provisions of the EMC Directive on being *placed on the market* before it is taken into service'. This comment appears to be nonsense as the product is invisible to the market. This was also discussed in 3.2.1. Article 10 of the EMC Directive only refers to the 'CE' marking in the context of 'placing on the market' and therefore for equipment *'taken into service'* there is no requirement to affix the 'CE' marking, although it is still required to meet the protection requirements. Perhaps this statement in the explanatory document is an attempt to ensure that all products are, at least technically, 'placed on the market' and therefore require to be marked with the 'CE' marking?

This particular aspect was considered earlier in the DTI's consultative document (1989), which suggested that where apparatus is built by the user and is not commercially available, certification of compliance is not required. Enforcement powers will be available against the user however (as permitted under Article 6 of the EMC Directive), if the apparatus is found not to comply or causes an actual or potential interference problem. A warning is given that 'it may be desirable' to carry out certification of some types of equipment before they are taken into service (even if they are not placed on the market) if they are likely to present a serious threat to EMC. The example of RF heating apparatus is cited.

Custom 'one-off' apparatus was also considered to come under the category of 'taken into service' by the consultative document. This would appear not to be the case since 'placed on the market' also covers the situation where a product is disposed of directly to the user.

3.2.3 Manufacturer

The 'manufacturer' is defined as the 'person' (organisation) accepting responsibility for the design and manufacture of a product falling within the scope of the EMC Directive.

The manufacturer may be established within or without the EEA. In either case he may appoint an 'authorised representative', more usually referred to as an agent, to act on his behalf and who must be established within the EEA.

The manufacturer is responsible for designing and manufacturing a product in accordance with the essential protection requirements laid down in Article 4 and Annex III of the EMC Directive. He is also responsible for following the certification procedures to demonstrate the conformity of the product with the requirements of the EMC Directive. The manufacturer is also permitted to sub-contract these tasks or use ready-made items or components and retain his position as 'manufacturer'.

The explanatory document gives three further examples of manufacturers:

- any 'person' who produces a new finished product utilising existing finished products is considered to be the manufacturer of the new product.
- any 'person' who modifies, transforms or adapts a product becomes the manufacturer of a new product. Beware the DIY electronics hobbyist, tampering with your Hi-Fi, it may make you a manufacturer!
- any 'person' who imports a used product from outside the EEA which is required to comply with the essential protection requirements of the EMC Directive, has the same responsibility as a manufacturer.

3.2.4 Authorised representative

The authorised representative (more commonly known as an agent) is appointed by a manufacturer to act on his behalf within the EEA. The EMC Directive places the same responsibility normally carried by a manufacturer onto an authorised representative to ensure that apparatus complies with the essential protection requirements.

3.2.5 Importer

An importer is a 'person' importing a product manufactured outside the EEA which is within the scope of the EMC Directive.

3.2.6 Final user or consumer

This is the user of a product placed on the market where the product does not need to be modified, adapted or fitted into any more complex apparatus. Presumably this latter constraint does not apply to computer peripherals. It may also be argued that a user fitting a plug-in card to a personal computer has now become a manufacturer, see 3.2.3!

Article 6 of the EMC Directive allows Member states to make provisions, for example to protect public telecommunications networks, and it is these local provisions which are likely to affect users directly.

3.3 Scope of the EMC Directive

In Chapter 2 the scope of the EMC Directive was described as encompassing all electrical and electronics equipment and all electromagnetic phenomena. It is clearly important for industrial sectors to know whether their products are covered by the EMC Directive. Here, the 'umbrella' approach whilst being 'a catch all' has generated much discussion and because of its all encompassing nature has inevitably produced the reaction 'does this really mean me?'. In practice the EMC Directive is quite clear and particular industrial sectors are listed in Annex III of the Directive. The explanatory document goes further in explaining what is covered.

Generally the structure of the explanatory document will be followed, but the treatment of other Directives having EMC provisions is somewhat confusing and will be dealt with in a different way.

3.3.1 Equipment covered by the EMC Directive

The following listing [EMC Directive[1], Annex III] is of equipment covered by the EMC Directive but should not be considered as 'restrictive'.
Emission and immunity aspects

- electrical household appliances, portable tools and similar equipment;
- fluorescent lighting luminaires fitted with starters;
- lights and fluorescent lamps;

- commercially available amateur radio equipment;
- radio and television receivers;
- industrial equipment; however what is meant by industrial equipment is not defined. Reference must be made to standards. CENELEC have defined the industrial environment in the context of the published industrial generic standards (*see* Chapter 6);
- radio and television broadcast transmitters;
- information technology equipment;
- aeronautical and marine radio apparatus;
- educational electronic equipment; the level of emitted disturbance may exceed the 'levels of the essential protection requirements' when equipment is intended for studying electromagnetic phenomena. However, the training, research or educational establishment is required to take all necessary measures to ensure that equipment outside the electromagnetic environment can function as intended;
- telecommunications apparatus.

3.3.2 Equipment excluded from the EMC Directive

Where Directives for specific product groups include EMC provisions either wholly or in part, then these will take precedence over the EMC Directive. If the provisions do not cover all the aspects of the EMC Directive, then there will be a default to the EMC Directive for those aspects. The Directives with EMC provisions are listed in Table 3.1.
Other exclusions emission and immunity aspects

- Radio equipment used by radio amateurs (unless the apparatus is commercially available). This exclusion has been included because of the nature of the activity which does not involve any kind of commercial transaction. 'Radio amateurs are persons carrying out *experimental activities* within the field of radio communications, according to definition No. 53 of the ITU Radio Regulation.'

3.3.3 Components

A 'component' is defined as an item which is used in the composition of an apparatus and which is not itself an apparatus with an intrinsic function intended for the final consumer.

Examples given in the explanatory document are: integrated circuits, electronic cards, minature resistors, or small capacitors. A component may be more complex as long as it does not have an intrinsic function and its only purpose is to be incorporated inside an apparatus.

On this basis components are *excluded* from the scope of the EMC Directive. However there is an onus on the component manufacturer to indicate to an equipment manufacturer how to use and incorporate a particular component. It is noted that if an additional component is added (or if a component is replaced by a replacement part) to an apparatus covered by the EMC Directive, the apparatus is still required to comply. The implications of this are discussed in Chapter 10 (IEEE Communications card).

Table 3.1 Directives with EMC provisions for specific product groups

Directive	*Emission/Immunity Aspect*	*Implementation Date / Transition Ends*	*EMC Directive*
Active Implantable medical devices 90/385/EEC	Emissions Immunity	01 July 1992 Transition ended 31 December 1992	No
Medical devices 93/42/EEC	Emissions Immunity	01 Jan 1995 Transition ends 13 June 98	No
Motor vehicles 72/245/EEC	Emissions	Revision to include full EMC under discussion No dates	Not expected to apply
Tractors & agricultural vehicles 72/322/EEC	Emissions	No published changes	Not expected to apply
Non-automatic weighing instruments 90/384/EEC	Immunity	01 Jan 93	Emissions aspect
Telecoms terminal equipment 91/263/EEC & 93/97/EEC	Emissions	06 Nov 92	Apply for provisions *not* covered by 91/263/ EEC
'EMC Directive' all other apparatus 89/336/EEC	Emissions Immunity	01 Jan 92 Transition ends 31 Dec 95	Compliance action plan see Figure 10.1

*Up to 13 June 1998 medical devices may: comply with existing national EMC regulations (for the UK the WT Regs.1963) or comply with the EMC Directive or the Medical Devices Directive and carry the CE marking. After 13 June 1998 Medical Devices must comply with the Medical Devices Directive [SI 1994 No. 3080].

3.3.4 Systems and installations

The explanatory document defines apparatus, equipment, a system and an installation, as follows:

- 'Apparatus' and 'equipment' are synonymous and mean a finished product having an intrinsic function, intended for the 'final user' and intended to be 'placed on the market' as a single commercial unit. Examples are a domestic sewing machine, an electric lawnmower, a bench power supply and an industrial vacuum cleaner.

- 'System' means several items of apparatus (or equipment) combined to fulfil a specific objective and intended to be 'placed on the market' as a single functional unit. The most obvious example is a personal computer consisting as a minimum system of keyboard, monitor, 'host' computer containing hard and floppy disk drives and CPU card, to which can be added a number of peripherals, such as a printer. Where apparatus which can be used in a system, is placed on the market, it should comply, *when used in a system*, so that the manufacturer complies with Article 3 of the Directive[1], which states that apparatus must be compliant 'when it is used for the purposes for which which it is intended'.

- 'Installation' means several combined items of apparatus or systems put together at a given location to fulfil a specific objective but not intended to be placed on the market as a single functional unit. An example would be a manufacturing facility consisting of: CNC machine tools, arc welding equipment, RF heat treatment equipment, a data logging system and a standby diesel generating set.

The explanatory document then considers how the EMC Directive applies to apparatus, systems and installations. The EMC Directive is clearly applicable to apparatus and to systems comprised of apparatus designed and intended to be operated together. An installation may be a random combination of apparatus and/or systems in all possible configurations. Each apparatus or system used in an installation is subject to the provisions of the EMC Directive. When used in an installation the apparatus and systems must meet the the manufacturer's installation guidelines and therefore should ensure interference- free operation of the installation itself and therefore the installation is *excluded* from the scope of the EMC Directive.

3.4 Electromagnetic Phenomena

The explanatory document simply restates Article 4 and the principal objectives set out in Annex III *ie*, the essential protection requirements (*see* Chapter 2). The emission limits and the levels of immunity to the different forms of electromagnetic disturbance are undefined by the EMC Directive itself. These and the methods of measurement are given in the harmonised standards. The explanatory document adds a wry note 'in view of the very general nature of the protection objectives, the proper operation of the EMC Directive depends to a large extent on the existence of standards'. As will become apparent from Chapter 5 there still remains a serious shortfall of standards related to immunity and large equipment/systems.

3.5 Procedures for Assessment of Conformity of Products Intended to be Placed on the Market

Article 10 of the Directive specifies three procedures for assessment of the conformity of apparatus:

- Article 10 (1) describes the procedure for apparatus to which the manufacturer can apply the harmonised standards.
- Article 10 (2) describes the procedure where the manufacturer has not applied the standards, only applied them in part, or in the absence of standards.
- Article 10 (5) describes the specific procedure for equipment designed for the radio transmission of radiocommunications.

3.5.1 The procedure for the assessment of conformity —Article 10 (1)

The procedure for assessment of conformity described by Article 10 (1) is *based on module A of Council Decision 90/683/EEC*[5]*, 13 December 1990.*

It is interesting to note that the EMC Directive makes no reference to 90/683/EEC and that the EMC Directive pre-dates it anyway! Presumably the European Commission was trying to tidy up the assessment procedures in all of the new approach Directives and took advantage of the explanatory document to link the EMC Directive with 90/683/EEC.

The explanatory document adds little to understanding the 'self-certification' route to compliance described in Chapter 2 sections 2.4.1 and 2.5. It does however state that the declaration of conformity (2.5) should not only include reference to the harmonised standards which have been applied but, where appropriate, also indicate details regarding the application of the standards and the results of tests which have been carried out. The manufacturer (or his authorised representative) keeps the declaration of conformity at the disposal of the relevant authorities for a period of 10 *years after the last product has been manufactured.*

The explanatory document also makes reference to the manufacturing process. A manufacturer should take all necessary measures to ensure that in the process of manufacture, products conform with the declaration of conformity, such that all products will be in compliance with the protection requirements of the EMC Directive. This means assuring the quality of production which may require the introduction of a quality assurance scheme such as BS 5750[6] (or ISO 9000). Many of the harmonised standards also include the '80/80 rule' defined in CISPR publication 16[7]. This rule requires that there is an 80% confidence that 80% of production will meet the limits specified, this is determined using statistical methods on the test results measured from a sample batch of products. Conformity of production may also require 'go/no-go' production testing to be devised.

3.5.2 The procedure for the assessment of conformity — Article 10 (2)

The 'Technical Construction File (TCF) route', Article 10 (2) is based on module A of Council Decision 90/683/EEC of 13 December 1990. Again this procedure is apparently based on a document post-dating the EMC Directive, *see* 3.5.1 above.

Essentially the explanatory document reiterates the description of the TCF route to compliance, see Chapter 2, section 2.4.2. It does however imply that the TCF should be used where harmonised standards cannot be used or are only applied in part. A technical report or certificate from a competent body must be obtained and the explanatory document states that either of these 'certifies only the conformity of the product with the essential protection requirements not covered by the harmonised standards'. The manufacturer or his authorised representative should keep with the TCF a copy of his declaration of conformity, this should be at the disposal of the relevant authorities for a period of 10 years after the last *product has been manufactured*. Where neither the manufacturer nor his representative are established within the Community, the TCF should be kept available by the person responsible for placing the product on the market.

The manufacturer should take appropriate measures to ensure that manufactured products conform to the TCF. As for self-certification this may involve operating a documented quality assurance scheme for the manufacturing process and perhaps some form of EMC evaluation during production testing.

3.5.3 The procedure for the assessment of conformity — Article 10 (5)

The procedure for assessment of conformity described by Article 10 (5) is based on modules B and C of Council Decision 90/683/EEC, 13 December

1990 (*see* comments in 3.5.1).

This assessment procedure is only concerned with 'apparatus designed for the transmission of radiocommunications, as defined in the ITU Convention'. A manufacturer is required to obtain an 'EC type-examination certificate' to demonstrate compliance with the provisions of the EMC Directive. This certificate is issued by a 'notified body'.

Module B describes the procedure followed by a notified body in order to 'ascertain and attest' that a production sample of the apparatus meets the provisions of a particular Directive.

A manufacturer (or his authorised representative) established within the Community, applies to a notified body for EC type-examination. This may be a notified body of the manufacturer's choice. The application includes:

- the name and address of the manufacturer (or authorised representative)
- a written declaration that the same application has not been lodged with any other notified body
- product technical documentation

The applicant provides the notified body with a representative specimen of production equipment for carrying out the test programme, denoted the 'type'. The notified body may request further samples. The notified body will then arrange to perform the appropriate examinations either in its own laboratories or it may sub-contract this work to an independent laboratory. The technical documentation, provided by the manufacturer, should enable the product to be assessed for compliance with the protection requirements of the Directive. This documentation should therefore cover the design, manufacture and operation of the product.

The notified body will:

- examine the technical documentation, verify that the type conforms with the manufacturing information and identify those 'components' which are designed in accordance with the Directive provisions and those which are not.
- perform (or have performed) the appropriate examinations and necessary tests to check whether the product meets the essential protection requirements of the Directive, where standards have not been applied.
- perform appropriate examinations, *etc.* to check that relevant standards have been applied, where these have been applied by the manufacturer.
- agree with the applicant the location where the examinations and tests are to be carried out.

Where the type meets the provisions of the Directive, the notified body issues an EC type examination certificate to the applicant. The certificate contains the manufacturer's name and address, conclusions of the examination, any conditions placed on the validity of the certificate and identification of the approved type. A list of the relevant parts of the technical documentation is annexed to the certificate and a copy kept by the notified body.

If a manufacturer is denied a type certification, then the notified body must provide detailed reasons. Provision must be made for an appeals procedure.

Should a manufacturer modify a product then the notified body should be informed of these modifications when conformity with the essential protection requirements is affected. Further approval is given by an addition to the original certificate.

The notified bodies communicate information on the EC type-examination certificates and additions issued or withdrawn, to to the other notified bodies.

The manufacturer or his authorised representative (or the person placing the product on the market if either the manufacturer or his representative are not established within the EEA) keeps with the technical documentation a copy of the EC type-examination certificate (and any additions) for a period of *10 years after the last product has been manufactured.*

Module C describes the procedure for manufacturers to ensure and declare that products are in conformity with the type for which an EC type-examination certificate is held and satisfy the protection requirements of the Directive. Essentially the manufacturer (or his authorised representative) affixes the CE marking and makes a declaration of conformity, which is held for a period of *10 years after the last product has been manufactured.* This is as previously described for both of the other routes to compliance.

A manufacturer should take appropriate measures to ensure that manufactured products conform to the type for which an EC type-examination certificate is held. As for self-certification and the TCF routes to compliance, this may involve operating a documented quality assurance scheme for the manufacturing process and perhaps some form of EMC evaluation during production testing.

3.5.4 Competent authorities, competent bodies and notified bodies

The EMC Directive refers to three bodies which exercise different functions. These are: competent authorities, competent bodies and notified bodies.

Competent authorities

The competent authorities are the national administrations of the Member

States responsible for fulfilling the responsibilities placed on them by the EMC Directive (Chapter 2). Each member state must notify the competent authorities to the European Commission and the other member states. The competent authority in the UK is the Department of Trade and Industry.

Competent bodies

Annex 2 to the EMC Directive sets out the criteria for competent body status. Organisations which conform to the EN 45 000[8] series of standards and which can provide documentary evidence to support this are presumed to be competent. Manufacturers' own test facilities may also be recognised as long as they fulfil the criteria and can prove their independence. It is not required to notify a competent body to the European Commission or the other member states, but bodies accepted as competent are advised 'to make themselves known to the Commission, which will publish their names and addresses'.

The function of the competent body is to issue technical reports or certificates for inclusion in a TCF.

A body is recognised as competent either by an accreditation body recognised by the competent authority, or by a body representing the competent authority of a member state.

The UK interpretation of a competent body requires, as promulgated in the consultative document and as largely put into practice, NAMAS accreditation for test laboratories, *see* Chapter 2, 2.4.2. This assumes NAMAS is assessing the technical performance of test houses and not just the bureaucracy surrounding the procedures, quality manual and reports! Currently NAMAS is similar to BS 5750[6] and is essentially a quality assurance scheme that ensures that testing to a particular standard is carried out in the prescribed manner and that the equipment used is in calibration. This does not necessarily align with the remit of the competent body which should have the technical expertise to assess an equipment when applicable standards do not exist and must therefore be capable of adapting recognised test methods, or devising measurements, which will demonstrate the equipment's ability to comply with the essential protection requirements.

At the beginning of 1992 several large companies had achieved NAMAS accreditation for their test facilities. One reason for this is so that they do not have to re-test existing equipment. They intend to issue TCFs based on the testing they have already carried out to in-house standards, which in many cases exceed the requirements of relevant standards and as a competent body they will be able to issue their own technical report.

If it is required to use a third party for *self-certification* testing then use of a NAMAS approved test facility will give confidence in the quality of the testing, but *NAMAS is not required* and it may well be less costly for a manufacturer to perform his own testing where possible or to use a non-accredited test house.

Notified bodies
A notified body according to the EMC Directive (and the explanatory document) is identical to a competent body, excepting that it can also issue EC type-examination certificates. Bodies responsible for issuing EC type-examination certificates must be notified by each member state to the Commission and the other member states. The UK regulations[9] define the UK notified bodies, as described in Chapter 4, but the DTI has chosen not to make competent and notified bodies synonymous.

The explanatory document states that a manufacturer always has the choice of contacting a notified body which has been notified in another member state. Presumably this is also true of competent bodies.

3.6 Safeguard Clause [Article 9]

The safeguard clause is the procedure taken when a Member State takes action against a product bearing the CE marking. The Commission must be immediately notified, there is then a consultation procedure between the Commission and the parties concerned. This is described in Chapter 2, section 2.5.

3.7 Progress on Standardisation

The final section of the explanatory document covers European standards. These are described in Chapter 5. Since the explanatory document was published a list of harmonised EMC standards has been published in the Official Journal, No C 44/12, 19 February 1992[10], the first two generic standards in the OJ, No C 90/2, 10 April 1992[11] and the generic emission standard for the industrial environment along with amendments to product standards in the OJ, No C 49/3, 17 February 1994[12].

3.8 Summary

The explanatory document clarifies the terminology used in the EMC Directive. It also clarifies some of the important questions which have been raised relating to the scope of the EMC Directive. Particularly the terms 'placed on the market' and 'taken into service' are explained and what is meant by apparatus (equipment), systems and installations. Conformity assessment procedures are described and particularly for radiocommunications transmitters the EC type-examination procedure is expanded when compared with the text of the EMC Directive.

References

1. 89/336/EEC Council Directive 'on the approximation of laws of Member States relating to electromagnetic compatibility', Official Journal of the European Communities No 139, 25 May 1989, pp19-26
2. 'Electrical Interference: a Consultative Document', DTI/PUB207/10k 10.89
3. EC explanatory document on Council Directive 89/336/EEC, 111/4060/91/EN-Rev. 1
4. 92/31/EEC Council Directive of 28 April 1992 'amending Directive 89/336/EEC on the approximation of laws of the Member States relating to electromagnetic compatibility', Official Journal of the European Communities No L 126/11, 12 May 1992
5. 90/683/EEC Council Decision of 13 December 1990, Official Journal of the European Communities No L 380, 31 December 1990.
6. BS 5750: Parts 0 to 4: 1987 (ISO 9000 and EN 29000) 'Quality management and quality systems', British Standards Institute, 1987
7. CISPR publication no.16 'CISPR specification for radio interference measuring apparatus and measurement methods', IEC, 1987
8. EN 45000 Series of standards, BS equivalent BS 7500 series, 'Criteria for the operation of testing laboratories', British Standard publications, 1989
9. 'Electromagnetic Compatibility (EMC) Draft United Kingdom Regulations - A Consultative Document', Department of Trade and Industry, July 1992
10. 92/C 44/10 Commission Communication in the framework of the implementation of the 'new approach' Directives 'publication of titles and references of European harmonised standards complying with the essential requirements', Official Journal of the European Communities No C 44/12, 19 February 1992
11. 92/C 90/02 'Commission Communication in the framework of the implementation of Council Directive No 89/336/EEC of 3 May 1989, in relation to electromagnetic compatibility', Official Journal of the European Communities No C 90/2, 10 April 1992
12. 94/C 49/3 'Commission Communication in the framework of the implementation of Council Directive No 89/336/EEC of 3 May 1989, as amended by Council Directive No 92/31/EEC, in relation to electromagnetic compatibility', Official Journal of the European Communities No C 49/3, 17 February 1994

4
Legislation

The EMC Directive must be implemented in each Member State of the European Community. This Chapter is principally concerned with the UK regulations. For UK companies or importers of equipment from outside the EEA resident in the UK, these regulations define the enforcement procedures adopted in the UK and the enforcement authorities. It should therefore be read by senior management in order to appreciate the potential penalties their companies may face if their equipment does not comply with the protection requirements of the EMC Directive. These regulations implement the Directive, the amending Directive and draw heavily on the explanatory document and therefore this chapter should be read in conjunction with Chapters 2 and 3.

4.1 Introduction

Under the original timetable for the EMC Directive [89/336/EEC[1]] the implementing legislation in each Member State of the European Community (EC) was to have been completed during the summer of 1991 and to have been effective from 01 January 1992. Only one Member State, Denmark, had implemented the EMC Directive by January 1992, the draft German legislation existed, but the remaining states, including the UK were awaiting the outcome of the European Parliament's legislative process with respect to the amending Directive [92/31/EEC[2]]. The amending Directive agreed on 28 April 1992 required each Member State to implement the amended EMC Directive by 28 July 1992 and for the legislation to become effective from 28 October 1992 (*see* Chapter 3).

During the transitional period the legislation existing in each Member State on 30 June 1992 operates in parallel with any legislation implementing the EMC Directive. These regulations are considered in Chapter 11. The

Germans published Verfugung (Vfg.) 241 on the 11 December 1991 implementing the transitional requirements, this was subsequently modified by Vfg. 46 in 1992 [EMC Europe, 1994[3]]. In the UK the transitional arrangements are included within the legislation which implements the EMC Directive.

This chapter will cover the UK legislation effective from 28 October 1992 and is based on the 'Electromagnetic Compatibility Regulations' Statutory Instrument (SI) 1992 No. 2372[10] and the amendment SI 1994 No. 3080[11]. The purpose of this chapter is to interpret the legislation such that the key issues affecting manufacturers of electrical and electronic equipment are clear. For the fine details of the legislation or precise wording the reader is referred to the Statutory Instruments themselves.

4.2 'The Electromagnetic Compatibility Regulations' SI 1992 No. 2372 — the UK Implementation of Directive 89/336/EEC

The regulations are arranged in eight parts. The first six parts transpose the amended EMC Directive into UK law and 'Part VII' is concerned with the UK enforcement of the regulations. The eight parts are as follows:

Part I Preliminary
Part II Application
Part III General requirements
Part IV The standards route to compliance
Part V The technical construction file route to compliance
Part VI The EC type-examination route to compliance for radio communication transmission apparatus
Part VII Enforcement
Part VIII Miscellaneous and supplemental

Each part will be considered in turn and for cross-reference puposes to the regulations, the regulation numbers will be used.

An amendment, SI 1994 No. 3080[11], removes some further areas open to misinterpretation but principally adds the requirements of the CE Marking Directive 93/68/EEC and recognises the extension of the EMC Directive to the European Economic Area (EEA). These additions have been incorporated where appropriate.

4.2.1 Part I Preliminary

1. The regulations are to be cited as the Electromagnetic Compatibility Regulations 1992 and came into force 28 October 1992.

2. Section 12A of the Wireless Telegraphy Act (WT Act) 1949[4] and section 78 of the Telecommunications Act (T Act) 1984[5] are repealed. (This provision allowed regulations to be made in respect of the immunity of radio equipment but has not been exercised, see Chapter 12.)

3. Regulation 3 is concerned with interpretation and defines terms used in the regulations. Many of these terms have already been defined in chapters 1, 2 and 3 and therefore only those terms which are specific to the understanding of the regulations will be defined where appropriate.

'Kits' are defined as a collection of all or substantially all the necessary components supplied as a single commercial unit, to construct an item of electrical apparatus, whether or not supplied with instructions. The amending regulations add that a kit may be a system with an intrinsic function for the end user which is supplied unassembled. (This marks a change in the DTI's understanding of the Directive when compared to the 1989 consultative document[5] which suggested that kits would be excluded and at the time was seen as a significant 'loophole'.)

A '*system*' is defined as an item of equipment, or combination of items of equipment which is comprised of electrical components, electronic components, or both (*see* 3.3.4).

Relevant apparatus (other than a kit) is *'taken into service'* when it is first used by the person who assembled it, or the person importing it from a country or territory other than a member state. However equipment operated for demonstration purposes or at a trade fair or exhibition is not considered to have been taken into service and therefore does not need to comply.

After 1 January 1994 reference to the 'Community' includes the 'European Economic Area (EEA)' and reference to a member state includes an EEA state.

4. Electromagnetic Disturbance is defined in general terms as any electromagnetic phenomenon which is liable to degrade the performance of 'relevant apparatus'. This is then expanded in line with Article 1.2 of the EMC Directive[1] to be: electromagnetic noise, unwanted signals and changes in the propagation medium. What an electromagnetic disturbance may be is listed in schedule 2 to the regulations:

i) Conducted low-frequency phenomena —

- harmonics, interharmonics
- signalling voltages
- voltage fluctuations
- voltage dips and interruptions
- voltage unbalance
- power-frequency variations
- induced low-frequency voltages
- dc in ac networks and

- dc ground circuits

ii) Radiated low-frequency phenomena —
- magnetic fields and electric fields

iii) Conducted high-frequency phenomena —
- induced continuous wave (CW) voltages or currents
- unidirectional transients
- oscillatory transients

iv) Radiated high frequency phenomena —
- magnetic fields
- electric fields
- electromagnetic fields
- continuous waves
- transients

v) Electrostatic discharge phenomena (ESD)

Therefore it can be seen that for regulatory purposes the DTI believes that more guidance should be available to manufacturers regarding what constitutes an EM disurbance than is given in either the Directive or the explanatory document[6].

Where equipment produces emissions as a function or consequence of its operation this is not considered to be an EM Disturbance if it is permitted and if it does not exceed limits specified by the W T Act[4]. Also for the purposes of the regulations a nuclear electromagnetic pulse (NEMP) is not regarded as a disturbance.

Surprisingly performance degradation criteria are defined in the regulations, intuitively it might be thought that these technical requirements should be left to the harmonised standards. Equipment is degraded if there is: a permanent, temporary or intermittent total loss of function, or there is a *significant* impairment of function, or in the case of equipment using a storage medium, destruction or corruption of stored information.

5. Finally Part I states the protection requirements as set out in the EMC Directive (*see* Chapter 2, 2.2). Particularly the regulations refer to the principal protection requirements of the EMC Directive[1] which are set out in Annex III to it. That is equipment must not generate an EM disturbance which will affect the equipment or services listed, when it is:

i) properly installed and maintained; and

ii) used for the purpose for which it was intended'. The listing of Annex III is reproduced in the regulations as schedule 3 and appears in this text in Chapter 3, 3.3.1.

Regulation 5 (5) makes three points in respect of equipment immunity requirements:

i) the immunity level should be what might be reasonably expected from the equipment consistent with its *function or intended function*

ii) the manufacturer should provide the final user with a specification for what is considered to be an *acceptable level* of degradation of performance

iii) the consequences of performance degradation: - the operation of apparatus may not be dangerous (either to persons or property) in any reasonably foreseeable circumstances

4.2.2 Part II Application

This part of the regulations is concerned with defining *'relevant apparatus'*. This is the term used to describe all equipment to which the regulations apply. It is also concerned with defining equipment which is excluded from the regulations. In essence therefore this part implements the scope of apparatus covered by the EMC Directive[1] as detailed in the explanatory document[4] and described in Chapter 3 (3.3).

Equipment covered by the regulations

6. & 7. For the purposes of these regulations relevant apparatus is defined as:
a) — 'a product with an intrinsic function intended for an end user'
b) — 'supplied or intended for supply or taken into service as a single commercial unit'. This unit may be:

i) an electrical appliance
ii) an electronic appliance
iii) a system

Modified Application

Some equipment is subject to a modified application of the regulations:

8. The application of the regulations to *educational electronic equipment* is defined as in the explanatory document, see 3.3.1. This regulation is modified by SI 1994 No. 3080 to refer to *Education and training equipment* which does not conform with the protection requirements in normal use in its usual EM environment. Such equipment is considered to meet the regulations if:

— it is accompanied by a declaration in English stating that use outside the classroom, laboratory, study area, etc invalidates conformity with the EMC Directive and could lead to prosecution.

— the equipment does not cause a disturbance to apparatus situated outside its immediate environment.

Education and training equipment is defined as any relevant apparatus or kit supplied to education and training establishments for experimentation, learning or practical training.

9. *Test apparatus* is also treated in the same way. This means that these types of equipment whilst generally conforming to the protection requirements are allowed to generate electromagnetic disturbances in a local environment provided that equipment immediately outside the environment is unaffected. This is good news for EMC test equipment manufacturers!

Equipment excluded from the regulations

10 &11. Transitional exclusions - these regulations do not apply to equipment supplied or taken into service in the EEA prior to 28 October 1992. These regulations also do not apply to any equipment for which compliance with the EMC Directive has not been claimed in the period up to 31 December 1995. However this latter group of equipment must comply with any national regulations existing in each member state prior to 30 June 1992.

General exclusions

12. Apparatus intended for *export outside the EEA* is excluded from the regulations, *ie* apparatus which will not be taken into service in the UK or in another member state.

13. *Excluded installations* are comprised of two or more combined items of relevant apparatus or systems put together at a given place to fulfil a specific objective but not designed by the manufacturer to be supplied as a single functional unit (*see* Chapter 3, 3.3.4). 'Manufacturer' is qualified to be 'manufacturers' where items are made by different manufacturers.

14. *Spare parts,* are defined as *components or combination of components*, intended for use in replacing parts of electrical apparatus, and are excluded from the regulations. However, apparatus into which a spare part has been incorporated is still subject to the regulations.

15. These regulations do not apply to *equipment supplied* by a manufacturer *to* his *authorised representative*.

16. *Second-hand apparatus* — equipment which has previously been taken into service, is excluded from the regulations. Unless, it has been subjected to further manufacture, this includes reconditioning or modification which substantially alters the EMC characteristics of the apparatus [regulation 3 (2)], or was previously supplied or taken into service outside the EEA.

17 *EM benign apparatus* — the regulations do not apply to apparatus inherently neither liable to cause nor be affected by an EM disturbance.

Specific exclusions

18 Apparatus for use in a sealed electromagnetic environment — equipment is excluded if it is intended to be used within a screened environment, it is accompanied by instructions which state that it is only for use in that environment and it is taken into service in that environment.

19 Radio amateur apparatus not available commercially is excluded.

20. Military equipment is excluded but not equipment which may be used for both military or other applications. (Note the French regulations include military equipment.)

21 to 27 Apparatus which is covered wholly or in part by other Directives these exclusions are described in Chapter 3, 3.3.2 and a list of the relevant Directives and their scope is given in Table 3.1.

4.2.3 Part III General requirements

28 to 30. The requirements for the supply and taking into service of apparatus is that:

a) the apparatus conforms with the protection requirements
b) the conformity assessment requirements have been complied with
c) the CE marking has been properly affixed
d) the manufacturer or his authorised representative has properly issued an EC declaration of conformity.

31 & 32. The *conformity assessment requirements* are:

a) the standards route to compliance (Part IV of the regulations)
b) the technical construction file route to compliance (Part V of the regulations)
c) the EC type-examination route in the case of radiocommunication transmission apparatus (Part VI of the regulations)

Where the conformity assessment requirements have been complied with then apparatus is presumed to comply with the protection requirements until the contrary is proved. The *presumption of conformity*.

33. The CE marking — the requirements for affixing the CE marking are as decribed by the EMC Directive[1] and in Chapter 2 (2.5). Note these requirements were modified by the CE Marking Directive 93/68/EEC and the amendment to the EMC regulations SI 1994 No. 3080. The regulations make it illegal for anyone to affix the CE marking or any other mark which may be confused with it unless the apparatus complies with the protection requirements and the conformity assessment procedures have been adhered to. The form of the CE marking is shown in Schedule 4 to the regulations. For apparatus already carrying the CE mark in the original form, which included the year compliance was claimed, it may continue to be supplied in this form up to 1 January 1997 after which it must comply with the amended regulations.

34. EC Declaration of conformity — the requirements for the EC declaration of conformity are as described in the EMC Directive[1] (Chapter 2, 2.5) and in section 4.2.4. The regulations also make it illegal to issue a declaration of conformity unless the apparatus complies with the protection requirements and the conformity assessment procedures have been followed.

A declaration may only be made for *relevant apparatus.*

35. Retention of documentation —the declaration of conformity must be retained (along with the technical construction file or type examination certificate, where appropriate) for a period of ten years beginning with the date on which the last item of relevant apparatus was supplied in the EEA.

4.2.4 Part IV The standards *(self-certification)* route to compliance

36 to 39 This route is defined in Article 10.1 of the EMC Directive[1] and is described in Chapter 2 (2.4.1) and Chapter 3 (3.5.1).

40. The form of the declaration of conformity is prescribed by this regulation. It shall:

a) be in English
b) give the name and address of the responsible person and of the manufacturer, where the two are not the same
c) be signed by or on behalf of the responsible person and identify that signatory
d) bear the issue date
e) give the particulars of the relevant apparatus sufficient to identify it
f) state the numbers and titles of the applicable harmonised EMC standards applied by the manufacturer
g) certify that the apparatus conforms with the protection requirements of the EMC Directive[1]

4.2.5 Part V The technical construction file route to compliance

The TCF route to compliance has been described in Chapter 2 (2.4.2) and Chapter 3 (3.5.2). Essentially the manufacturer will choose this route when there are no relevant harmonised standards applicable to his product or they can only be applied in part. The manufacturer is required to prepare a TCF and obtain a certificate or technical report from a competent body which certifies conformity of the product with the essential protection requirements. The manufacturer then makes his declaration of conformity (which is retained along with the TCF for a period of ten years after the last product has been manufactured) and affixes the CE marking. The regulations clarify and detail this procedure.

42. The TCF route must be followed when:

a) the manufacturer chooses not to apply the applicable harmonised standards. *Note* the explanatory document implies that this is a non preferred choice (*see* 3.5.4)

b) there is no applicable standard

c) there are applicable standards but these do not make complete provision for the EM characteristics of the apparatus in order to satisfy the protection requirements.

Radiocommunication transmission apparatus is excluded from the TCF route.

43. - (1) and (3) The TCF should:

a) describe the apparatus (sufficiently to enable the enforcement authority to identify it)

b) contain information about the design, manufacture and operation of the apparatus

c) set out the procedures used to ensure conformity with the protection requirements for all the EMC characteristics of the apparatus (to enable the enforcement authority to ascertain whether it conforms with the protection requirements) or those for which a harmonised standard has not been applied

d) include a technical report or certificate

e) exist in material (hardcopy) form

f) give the name and address of the responsible person and of the manufacturer, where the two are not the same

g) state the numbers and titles of any EMC standards, specifications or codes of practice applied by the manufacturer

g) specify the electromagnetic environment for which the apparatus is suitable

(2) The apparatus for which the TCF may be appropriate is clarified by the regulations:

a) a single item of apparatus

b) a specimen production representative where a number of items are to be produced

c) a number of items of apparatus or representatives where each is a variant of the basic design

45. Competent bodies — a body responsible for issuing technical reports or certificates for inclusion in the TCF as described under Article 10.2 of the EMC Directive. Such a body should satisfy the requirements of Annex II to the EMC Directive and is appointed by a Member State. A competent body may be appointed to assess TCFs for all descriptions of relevant apparatus or for specific types of apparatus.

46. Appointment of UK competent bodies — a competent body is appointed in writing by the Secretary of State and authorised to act (as required by the EMC Directive) in respect of certain types of relevant apparatus. The types of apparatus which a competent body will be authorised to assess, will depend on the expertise and capability of the body. The appointment may be for 'the time being' or a defined period. Lists of the UK competent bodies will be published from time to time indicating the types of apparatus each body is authorised to evaluate and any conditions imposed on their appointment.

47 Eligibility and verification — the criteria for competent body status are listed in Annex II to the Directive and included within the regulations as Schedule 5. These are regarded as the 'minimum criteria' and a body is required to meet these for every type of apparatus for which it is appointed to assess. These minimum criteria are as follows:

i) availability of personnel and of the necessary means and equipment

ii) technical competence and professional integrity of personnel

iii) independence, in carrying out tests, preparing the reports, issuing the certificates and performing the verification function provided for in the EMC Directive, of staff and technical personnel in relation to all circles, groups or persons directly or indirectly concerned with the product in question

iv) maintenance of professional secrecy by personnel

v) possession of civil liability insurance unless such liability is covered by the government of the UK

(4) A body complying with the assessment criteria fixed by a harmonised standard is assumed to comply with the 'minimum criteria'. Such standards are the EN 45 000[7] series as indicated in the explanatory document, see Chapter 3 (3.5.4).

(5) This part of regulation 47 allows a manufacturer's testing laboratory to be eligible as a competent body. Essentially the laboratory must meet the minimum criteria, particularly demonstrating its independence from the rest of the company and a condition of appointment is that confidentiality of information is maintained by not disclosing such information to the rest of the company.

48. *Termination of appointment* — the appointment of a competent body may be terminated in writing by the Secretary of State if a body no longer satisfies the criteria or if it is necessary in the interests of manufacturers or final users! At termination provision is made for work in hand to be passed to another competent body.

49. *Competent body fees* — a competent body is permitted to charge for the preparation of technical reports or certificates as it chooses. This should be at the 'commercial rate normally charged (including profit)', an estimate should be provided to the applicant and the competent body is required to publish from time to time a scale of fees.

50. *Applications for technical reports and technical certificates* — a competent body is required to determine whether a technical report or certificate can be issued for an item of apparatus it is authorised to assess. It should consider the electromagnetic environment for which the apparatus is intended and any standards, whether harmonised or not, or any other technical criteria the competent body considers to be relevant. Where a competent body is of the opinion that the apparatus conforms to the protection requirements it will issue a technical report or certificate. Where it is considered that the equipment does not conform the application will be refused and reasons in writing provided.

51 For a competent body to accept an application for a technical report or certificate for 'relevant apparatus', the application must meet the following criteria:

(a) it must be in writing
(b) it should include a draft TCF which contains all the information which constitutes a TCF excepting the technical report or certificate
(c) it must include particulars of the EMC standards the manufacturer proposes to apply and the EMC characteristics to which they relate
(d) it must declare that no application for the apparatus has been made to another competent body which is outstanding

A competent body is only allowed to accept an application if it can produce the report or certificate within a three month period from receipt of the application. During the assessment a manufacturer is required to grant access to production facilities for the apparatus when reasonably requested by the body and the manufacturer should make available all information which may reasonably be required.

52 Contractors — a competent body is permitted to sub-contract testing, assessment or inspection on its behalf, or 'require the applicant to satisfy another person with respect to any matter'. The competent body however is not authorised to rely on the opinion of another as to whether apparatus conforms with the protection requirements. The contractor is permitted to charge a fee for the work he carries out.

53. Technical reports and technical certificates — a technical report or certificate is required to:

i) be in English
ii) give the name and address of the applicant and the manufacturer where the two are not the same
iii) be signed on behalf of the competent body and identify the signatory
iv) carry the date of issue and the number of the report/certificate
v) give the particulars of the relevant apparatus (and variants of it) sufficient to identify it and state whether the apparatus is a single item or a representative or a number of variants
vi) certify that the apparatus conforms with the protection requirements of the EMC Directive.

54. Conditions of technical reports or certificates — a competent body may issue a report/certificate conditionally or unconditionally as it considers appropriate. For example there may be a limitation on the environment for which the apparatus is suitable or the apparatus may be limited to an installation at a specific site. These conditions may be varied by the competent body which issued the report/certificate, new conditions may be imposed or conditions may be removed.

55. Withdrawl of technical reports or technical certificates — provision is made for a competent body to withdraw a report/certificate if subsequent to its issue it becomes apparent that the apparatus does not conform to the protection requirements. The person issued with the certificate must be informed in writing along with the reasons for the action.

56. Should a competent body refuse to grant a technical report or certificate, impose conditions, or withdraws the technical report or certificate, it must inform the applicant in writing along with its reasons. The applicant then has 28 days to discuss the decision with the competent body.

57. Declaration of conformity where compliance is shown solely using a TCF — the declaration is identical to that used for the standards route (see 4.2.4), but instead of listing harmonised standards the TCF is identified, the competent body issuing the technical report/certificate and the date and number of the report/certificate.

4.2.6 Part VI EC type-examination route to compliance for radiocommunication transmission apparatus

59 to 63. This route is applicable to 'wireless telegraphy apparatus' used for transmitting (transmitter) or transmitting and receiving (transceiver) other than radio amateur apparatus. This procedure for the assessment of conformity was described in Chapter 3 (3.5.3 and 3.5.4). Essentially an EC type-examination certificate is required to be issued for this type of apparatus by a body notified to the European Commission, a notified body.

The UK notified bodies authorised for particular types of apparatus are:

i) the Defence Research Agency of the Ministry of Defence (DRA)
ii) the Civil Aviation Authority (CAA)
iii) the British Approvals Board for Telecommunications (BABT)
iv) the Radiocommunications Agency of the Department of Trade and Industry (RA)

The descriptions of the apparatus for which each notified body is responsible are listed in Schedule 6 of the regulations and the reader is referred to this.

64 to 70. The applications for an EC type-examination procedure are handled by the regulations in a similar way to the TCF route. For example conditions may be placed on the use of the apparatus and the notified bodies may sub-contract the assessment of the apparatus. Finally it is made clear that nothing in the EMC regulations affects the licensing provisions of the WT Acts[4], the T Act[5] or the 1990 Broadcasting Act[8].

4.2.7 Part VII Enforcement

This part of the regulations details how the EMC Directive is to be enforced in the UK. It defines the enforcement authorities, procurement of test purchases, the powers granted to an enforcement officer (search, seizure of apparatus), prohibition and suspension notices, offences, misuse of the CE marking, penalties, power of the court and recovery of enforcement expenses.

73 Enforcement authorities — the enforcement authorities for relevant apparatus other than that specifically defined, are:

i) in Great Britain, the weights and measures authorities; and
ii) in Northern Ireland, the Department of Economic Development
iii) the Secretary of State

Enforcement for specific apparatus is delegated to:

iv) the CAA for Wireless Telegraphy apparatus as defined in paragraph 2 of Schedule 6 of the regulations
v) the Director General of Electricity Supply, in either Great Britain or Northern Ireland, for electricity meters

74. Test purchases — an enforcement authority or an officer of the authority, has the power to purchase electrical apparatus for the purpose of ascertaining whether it complies with the protection requirements. This means having the equipment tested.

75 & 76. Powers of search — provision is made for an officer of an enforcement authority to have power of entry to any premises (other than a person's residence) on the production of credentials, to inspect any electrical apparatus and examine any procedure connected with the production of the apparatus. The officer may also obtain a warrant from a Justice of the Peace (JP), allowing forcible entry, where there are reasonable grounds for suspecting contravention. If an officer has reasonable grounds for suspecting that there has been a contravention of the regulations he may seize and detain any electrical apparatus and any relevant documentation, if he has reasonable grounds for believing they may be required as evidence or are liable to be forfeit. When equipment is seized, the person from whom it is seized must be notified, including importers of apparatus. The seized apparatus may be submitted to a test to determine whether it is in contravention of the regulations.

77. An *appeal* procedure is established with the provision for *compensation* for any loss or damage.

78 Prohibition notices: — the Secretary of State is empowered to issue 'prohibition notices' on suppliers or end users to prohibit relevant equipment considered not to comply with the EMC requirements from being supplied or used.

The prohibition procedure is set out in Schedule 7 to the regulations: an appeal may be made against a Prohibition Notice which must be considered by the Secretary of State, either the notice will be revoked or a person appointed to consider the representations. The appointed person will then consider the representations, may call and examine witnesses and will report to the Secretary of State. The Secretary of State will consider the report and decide whether or not to vary or revoke the notice.

79 & 80. Suspension notice: — if an enforcement authority has reasonable grounds for suspecting a breach of the legislation, then it can issue a 'suspension notice' prohibiting a person from supplying or using equipment for up to six months. Again there is an appropriate appeal procedure.

81. Documentation: — an officer of an enforcement authority must be given access to the appropriate documentation, declaration of conformity, TCF, EC Type examination certificate and may take copies.

Offences:

The Directive requires member states to take 'all appropriate measures' to ensure compliant apparatus is placed on the market or taken into service. This has been interpreted as requiring national legislation having adequate sanctions for breaches of the Directive's EMC requirements.

82 to 87. Offences — the following offences are included in the regulations:

82. knowingly supplying or taking into service relevant apparatus contravening the regulations (28 and 29)

83. contravention of a prohibition or suspension notice

84. provision of false or misleading information in the required documentation

85. knowingly affixing the CE marking or an inscription which may be confused with it, to non-compliant apparatus and/or issuing a declaration of conformity for that apparatus

86. (1) failure to assist an enforcement officer (obstruction)
(2) any person pretending to be an enforcement officer

87. failure to retain full documentation

***88. Defence of due diligence* — an admissible defence will be to demonstrate that a person has taken 'all reasonable steps and exercised all due diligence to avoid committing the offence'.**

89. Liability of persons other than the principal offender — where a person commits an offence due to the default of another person, then the second person is guilty of the offence and may be proceeded against, whether or not proceedings are taken against the original offender. In the case of a corporate body where an act of default can be attributed to the neglect of an individual (manager, director, secretary, *etc*) then the individual as well as the corporate body is guilty of the offence and is liable to be proceeded against.

90 & 91. — Proceedings may be taken within 3 years of the date of an offence or within 1 year of an enforcement authority having sufficient evidence to proceed whichever is the earlier. A court may also infer that apparatus did not comply at the time of being placed on the market or being taken into service if it is subsequently proven not to comply.

92. Penalties — a person found guilty of offences 83, 84 or 86(2) is liable on conviction to: imprisonment for up to 3 months, or a fine up to level 5 on the standard scale (£5000 on 1 July 1992), or both! A person guilty of 82, 85, 86(1) or 87 is liable to a fine up to level 5 only.

93. Remedial action — where a person is convicted under offences 82 or 85 and it is the court's opinion that the person can take remedial action in order for equipment to become compliant, the court may in addition to or instead of imposing punishment, order the person to take remedial action within a specified period of time. This period may be extended by order of the court.

94 & 95 Forfeiture — an enforcement authority may apply under this regulation for the forfeiture of any relevant apparatus contravening the regulations. An application may be made to a magistrates court where proceedings have been brought against a person committing either offences 82, 83, or 85. The application for forfeiture may be for some or all of the apparatus. The magistrates court will only grant a forfeiture order if it is satisfied that the equipment does not satisfy the protection requirements.

Forfeited apparatus will either be destroyed, disposed of for reconditioning or disposed of for scrap as directed by the court.

The 'person' being forced to forfeit the apparatus has the right of appeal to the Crown Court (County Court in Northern Ireland).

The procedure under regulation 95. takes account of the differences between English and Scottish law.

96. Recovery of enforcement expenses — the court may order a convicted person to reimburse an enforcement authority for any expenditure which has been incurred in connection with the seizure, detention or forfeiture of non-compliant relevant apparatus.

4.2.8 Part VIII Miscellaneous and supplemental

This part, regulations 97 to 101, is essentially administrative. Of note is regulation 97 which places restrictions on the disclosure of information by officers of the enforcement authorities or other persons involved in the enforcement procedures. A fine up to level 5 may be imposed or imprisonment up to 2 years or both! Therefore the potential penalty for a wayward enforcement officer is greater than for an unscrupulous manufacturer (excepting for the forfeiture of apparatus)!

4.2.9 Conclusions

The UK legislation is essentially in two parts. The first part encompassing Parts I to VI of the regulations transposes the EMC Directive and the amending Directive into UK law and implements the conformance assessment procedures. The second part, Parts VII (and VIII), is the UK enforcement procedure.

Whilst not explicitly stated in the regulations the enforcement authorities will be complaint driven, although the power is given to the authorities to purchase equipment and seize equipment

The penalties are not severe, but the potential forfeiture of apparatus and the prohibition or suspension from the market place are likely to be much more of a deterent to ensure that manufacturers comply with the protection requirements.

4.3 Legislation in other EC Member states

At the start of 1992 only Denmark had established laws implementing the EMC Directive, whilst Germany had circulated draft proposals.

4.3.1 Denmark

The Danish Electromagnetic Disturbance (Protection) Act, 10 Apri 1991. Lov Nr. 216 encompasses the laws existing on 30 December 1991 and describes the transitional arrangements, this came into force on 01 January 1992. Order No. 796 implements the EMC Directive [EMC Europe, 1994[3]].

4.3.2 Germany

The EMVG (EMC Law), 9 November 1992, implements the EMC Directive. The Federal Office for Post and Telecommunications is empowered to implement the law and the administrative duties are defined. The ZZF (*see* Chapter 12) is the designated notified body

The enforcement is to be paid for by raising a levy from transmitter operators in Germany. The estimated cost of implementing the law is 200,000,000 DM. It is anticipated that 5% of costs will be met by penalties and charges payable by manufacturers of non-compliant apparatus. Offences against the EMC Law are subject to fines of between 10,000 DM and 100,000

DM [Mertel, 1992[9]]. Offences are similar to those described for the draft UK regulations.

For updates on this information and details of the implementation in other countries both EU and EFTA the reader is referred to the BSi Technical Help to Exporters publication, 'Electromagnetic Compatibility Europe' ISBN 0 580 20917 2.

References

1. 89/336/EEC Council Directive 'on the approximation of laws of Member States relating to electromagnetic compatibility', Official Journal of the European Communities No 139, 25 May 1989, pp 19-26
2. 92/31/EEC Council Directive of 28 April 1992 'amending Directive 89/336/EEC on the approximation of laws of the Member States relating to electromagnetic compatibility', Official Journal of the European Communities No L 126/11, 12 May 1992
3. Technical Help to Exporters 'Electromagnetic Compatibility Europe', BSI, Issue 6 1994 ISBN 0 580 20917 2
4. Wireless Telegraphy Acts 1949 and 1967
5. Telecommunications Act 1984
6. EC explanatory document on Council Directive 89/336/EEC, 111/4060/91/EN-Rev. 1
7. EN 45 000 Series of standards, BS equivalent BS 7500 series, 'Criteria for the operation of testing laboratories', British Standard publications, 1989
8. Broadcasting Act 1990
9. H K Mertel 'European Community EMC Directives for Information Technology Equipment', ITEM 1992, R&B Enterprises.
10. Statutory Instruments 1992 No. 2372 'The Electromagnetic Compatibility Regulations 1994, HMSO, October 1992
11. Statutory Instruments 1994 No. 3080 'The Electromagnetic Compatibility (Amendment) Regulations 1994', HMSO, December 1994

5

A Guide to Relevant Standards

Harmonised standards are crucial to the implementation of the EMC Directive. It is the harmonised standards which specify the emission limits and immunity levels which equipment is required to meet. Approvals engineers and design engineers need to be familiar with these standards. By demonstrating compliance with the requirements of the harmonised standards manufacturers are able to self-certify that their products conform with the EMC Directive. This chapter defines what is meant by relevant or harmonised standards, product specific standards and generic standards. A table of equivalent standards and a user's listing of harmonised standards is provided. The shortage of product specific standards, immunity standards and standards for large installations is discussed.

5.1 Introduction

The EMC Directive [89/336/EEC[1]] defines two methods for demonstrating compliance with the objectives set out in Article 4. For self-certification, the manufacturer is able to declare that apparatus conforms to relevant standards. Alternatively, a technical construction file can be prepared. The latter must include a technical report, or a certificate from a competent body.

We are concerned in this chapter with the ***self-certification route*** to compliance.

For manufacturers to self-certify a product, it must be designed, built and tested to meet the requirements of *'relevant standards'*. A 'relevant standard' is defined by Article 7 of the EMC Directive as a national standard which has been *harmonised* with a standard whose reference number has been published in the Official Journal of the European Communities. In practice this means that a relevant standard is a *Euro Norm*, published by

CENELEC, the European Committee for Electrotechnical Standardisation.

The European Telecommunications Standards Institute (ETSI) is publishing EMC standards specifically for telecommunications equipment. It is possible that some of these may be relevant to other types of equipment for the purpose of demonstrating compliance with the EMC Directive.

Euro Norms (EN) are derived from CISPR and other IEC publications. It is necessary for individual EC member states to harmonise their own national standards with the appropriate EN. This means that identical standards will be used in all EC Countries. For example the British standard which covers emissions from Information Technology equipment is BS 6527. This is harmonised with Euro Norm 55 022.

There are two categories of relevant standard: the *product or product family specific standard* and the *generic standard.* A product specific standard applies to a particular type of product or group of products, for example EN 55 022 which applies to information technology equipment. A generic standard is categorised according to environmental type, for example 'residential, commercial and light industry' and applies to a broad range of product types.

Most CENELEC standards are derived from IEC standards or recommendations. The IEC committee responsible for EMC matters is CISPR. CISPR recommendations and other IEC documents which form the basis of a CENELEC standard can be identified in the standard's reference number. For example, EN 55 022 is derived from CISPR 22.

Table 5.1 provides a cross-reference showing the original CISPR document, the CENELEC Euro Norm and the equivalent British standard. It also shows the US and German approximate equivalents with which some electrical and electronic manufacturers' equipment may conform. These are discussed in detail in Chapter 12. When relevant standards are listed, Tables 5.2 to 5.5, it is clear that very few product types have been covered to date (May 1995).

Table 5.1 Equivalent Standards

Subject	USA	Germany VDE/DIN	British Standards BS	CISPR Publication	CENELEC
Emissions					
ISM	47CFR Pt.18	0871	EN 55 011	11	55 011
Ignition		0879	833	12	55012
Radio & TV	47CFR Pt.15	0872	905-1	13	55 013
Household appliances		0875-1	EN 55 014	14	55 014
Luminaires		0875-2	EN 55 015	15	55 015
Information technology	47CFR Pt.15	0878	6527	22	55022

Immunity		BS	CISPR	CENELEC
Radio & TV		905-2		55 020
Industrial-process measurement & control	SAMA PMC33.1 ANSI C63.12	BS 6667	IEC 801	HD481

Note: HD is a Harmonisation Document

5.2 Product-Specific Emission Standards

Product specific emission standards are listed in Table 5.2.

The radiated and conducted emission limits for 'Industrial, Scientific and Medical' (ISM) apparatus, is given in EN 55 011. This standard covers products which are radio frequency powered — for example woodglueing machines, induction heating equipment, and the CO_2 laser equipment used in hospitals. This standard is also referenced by the generic emission standard for the industrial environment, EN 50 081-2.

Other products covered are: broadcast receivers - radios and televisions; household appliances; luminaires, and information technology equipment (ITE).

EN 55 022 defines conducted and radiated emission limits for information technology equipment — this includes personal computers, printers, disk drives and many other peripheral products. This standard is referenced by the generic emission standard EN 50 081-1.

EN 60 555-2, defines the permitted harmonic content of mains disturbances caused by domestic apparatus. EN 60 555-3 defines permitted voltage deviation or 'flicker'. This standard is based on IEC 555 which is being revised and is likely to affect a wider range of electrical and electronic products which may be connected to a low voltage mains electricity supply. This is also referenced by EN 50 081-1.

5.3 Product-Specific Immunity Standards

When immunity standards are considered (Table 5.3) only one specific product type is catered for: broadcast receivers. IEC 801 for industrial process measurement and control equipment is included here as it is referenced by the generic immunity standards.

This shortfall in product-specific standards has been recognised by CENELEC which has introduced generic standards. Since the generic standards cover environments and therefore a broad range of product types, they will enable as many manufacturers as possible to self-certify their products.

5.4 Generic Emission Standard

The generic emission standard is EN 50 081. Part 1 covers 'residential, commercial and light industry' environments. Part 2 covers the 'industrial' environment.

Part 1 principally restates the emission limits and test methods defined by EN 55 022, which is the product specific emission standard for IT equipment. Part 2 does the same with EN 55 011, which is the product specific standard for ISM (Industrial Scientific and Medical) equipment.

5.5 Generic Immunity Standard

The generic immunity standard is EN 50 082. Parts 1 and 2 have the same environmental classification as the generic emission standard. This standard is largely based on IEC 801 which was originally established for industrial process measurement and control equipment. Part 2 is in draft form.

Generally, both product-specific and generic standards not only define emission limits and immunity levels, but also specify the test methods to be employed. Manufacturers using these standards to demonstrate compliance with the EMC Directive must be familiar with the contents, and appreciate the implications, of all relevant standards. In particular they must be aware of sections open to mis-interpretation, of deficiencies within the standards, and of test methods that require significant financial investment. Examples of standards illustrating these points are considered in detail in Chapters 6 and 7.

5.6 The Rate of Standards Development

The standards making machinery has been extremely active in an attempt to meet the deadlines imposed by the EMC Directive. The effect of the Directive has not only been upon CENELEC, but also upon the IEC which has found it necessary to revise CISPR recommendations and IEC 801. The pressures upon these bodies has resulted in closer co-operation to speed up the standards making process. This is described in Chapter 1 (1.3).

CENELEC has also established a fast route to producing a standard, this is known as the 'Vilamoura' procedure. Under this procedure a final draft standard should be produced in less than eighteen months. This procedure resulted from a proposal made at the 23rd General Assembly held in Vilamoura in May 1988 [BSI January 1991[2]].

In essence this procedure delegates the drafting of a particular European standard to the interested national standards bodies. The scheme requires the relevant national committee (the BSI EMC committee is GEL110) to announce all the national standards work being undertaken in a particular electrotechnical area. Other national committees can then register interest in this area of work. This results in the initial national proposal being converted into a European proposal. Work will then be commenced either under the auspices of a CENELEC Technical Committee or by the interested national committees. The latter method involves two or three interested national committees working together to provide the text of a draft European standard (EN), this requires no CENELEC resource and the timetable can be arranged to meet urgent market requirements. The resulting draft is sent to the national committees for formal vote in the normal CENELEC way (see Chapter 1, 1.3).

There were some 75 proposals across the whole electro-technical area being discussed by interested national committees in January 1991 [BSI January 1991[2]]. All of these were expected to produce ENs in short timescales with minimal use of central CENELEC resources while maintaining the rights of all members to be included in the approval procedures.

Despite the efforts of the IEC and CENELEC only one specific product type is covered by ENs for both emissions and immunity: broadcast receivers — both radio and television. Other equipment types must rely on the generic standards or a combination of product-specific and generic standards. Now that the generic standards have demonstrated the principle of calling up a number of 'basic' standards, it is to be expected that the rate of development of product and product family-specific standards will improve. New standards will also be able to reference basic standards and include additional test methods and set emission limits or immunity criteria specific to the particular product under consideration. An example of this is the draft standard for 'audio, video, audiovisual and lighting control equipment for professional use' which has been prepared by a British Standards Institute Panel Committee EEL/32/-/3 [Woodgate 1992[3]]. This committee reports to GEL110 which has submitted this draft to CENELEC under the Vilamoura procedure. Ratification is still awaited (April 1995).

5.7 List of Relevant Standards

For manufacturers requiring to self-certify their products for compliance with the EMC Directive a list of the available product-specific and generic standards is essential. Tables 5.2 to 5.5 represent the position in April 1995.

Table 5.2 Product Specific Emission Standards

EN 50065-1	BS EN 50065-1	Signalling on low voltage installations
EN 55 011	BS EN 55011	Industrial Scientific and Medical
EN 55 013	BS 905-1	Broadcast receivers & associated equipment
EN 55 014	BS EN 55014	Household appliances
EN 55 015	BS EN 55015	Luminaires
EN 55 022	BS 6527	Information technology equipment
EN 60 555	BS 5406	Limitation of disturbances in electricity supply networks caused by domestic and similar appliances equipped with electronic devices

Table 5.3 Product Specific Immunity Standards

EN 55 020	BS 905-2	Broadcast Receivers & Associated Equipment
Reference standards:		
HD481(IEC801pts 1,2,3)		Industrial Process, Measurement and Control
	IEC 801-1}	General
	IEC 801-2}BS 6667	ESD
	IEC 801-3}	Radiated Immunity
	IEC 801-4	Electrical Fast Transient/Burst

Table 5.4 Generic Emission Standards

EN 50 081-1	Generic Class: residential, commercial & light industry
EN 50 081-2	Generic Class: industrial

Table 5.5 Generic Immunity Standards

EN 50 082-1	Generic Class: residential, commercial & light industry
prEN 50 082-2	Generic Class: industrial

In addition to the European EMC standards listed in the OJ which may be used to demonstrate compliance with the EMC Directive via the standards route, CENELEC have adopted other product specific and basic standards. EN 60601-1-2: 1993 is for Medical Electrical Equipment and covers the EMC requirements within the general safety requirements. Complying with this standard enables medical electrical equipment manufacturers to comply with the EMC requirements of the Medical Devices Directive, 93/42/EEC. This standard is based on IEC 601-1-2. IEC 801-2 has been adopted as EN 60801-2 for industrial process measurement and control equipment. The IEC has also issued a number of basic standards in the IEC 1000 series, these have been adopted by CENELEC as the EN 61000 series and are a useful resource to those complying via the TCF route.

It would be impractical to review all of the relevant standards in detail, so EN 55 022 for information technology equipment, has been selected as an example of a product specific standard and EN 50 081-1 and EN 50 082-1 as examples of the generic standards (also included are the *basic standards* referred to in these generic standards). These are considered in Chapters 6 and 7 and are used to illustrate the deficiencies and implications of the standards for manufacturers who have chosen the self-certification route to compliance with the EMC Directive.

5.8 Conclusions

In conclusion it can be seen that very few product types are specifically covered by relevant standards. More emission standards exist than immunity standards and the only products currently covered by both product specific emission and immunity standards are sound and television broadcast receivers and associated equipment.

A large number of product types will be covered by virtue of the generic standards and at the present time it may be necessary for a manufacturer to use a combination of product specific and generic standards in order to self-certify his equipment as being in compliance with the EMC Directive. For manufacturers of physically large equipment such equipment is only partly covered within the ISM and ITE standards and therefore the technical construction file is likely to be the only option. This is discussed in more detail in Chapter 10.

The streamlining of the standards making procedures which has evolved within the IEC and CENELEC was expected to expedite the development of EMC standards during the EMC Directive's transitional period. In particular parallel voting in the two organisations (*see* Chapter 1, 1.3), the 'basic' standard concept and the Vilamoura procedure should have all contributed to the rapid provision of further product and product specific standards, however the position is little changed from that which existed in 1989!

References

1. Council Directive of 3 May 1989 on the approximation of the laws of the Member States relating to electromagnetic compatibility (EMC), 89/336/EEC, OJ L139 of 23.05.89, pp 19-26
2. '1992 BSI in EUROPE' (published with BSI News), January 1991, Issue 12
3. Woodgate, 'An industry-specific EMC standard', Audio Engineering Society - EMC Practicalities Conference, London, 13 May 1992

6

Interpreting Emission Standards

Chapter 6 is principally aimed at introducing the harmonised emission standards to technical managers, design engineers, EMC test engineers and approvals engineers. The implications in terms of costs of testing and test facilities are also applicable to those controlling finances or marketing within manufacturing companies. An overview is given of examples of product specific emission standards EN 55 022, and EN 55 011, which are also referenced by the generic standard EN 50 081-1 and 2. Also included is a review of the parts of other standards referenced by the generic standard. The problems of test repeatability are identified. Hence the reader will aquire an understanding of emission evaluation methods for radiated and conducted interference and an appreciation of the difficulties of repeatability involved for these types of test.

Harmonised or 'relevant' standards are crucial to the implementation of the EMC Directive. For products conforming with relevant standards compliance can be claimed with the EMC Directive [89/336/EEC[1]]. This route to compliance allows manufacturers to self-certify their products as described in Chapter 2. In this chapter examples of relevant emission standards are considered in some detail.

The standards considered in this chapter are:

i) EN 55 022[2] the 'product specific' standard which applies to Information Technology Equipment (ITE).

ii) EN 50 081-1[3].the generic emission standard for the 'residential, commercial and light industry' environment. This standard defines emission limits and references 'basic' standards for measurement methods; an overview of these basic standards is also given.

iii) the generic 'industrial' environment emission standard EN 50 081-2[4] and the basic standards referenced by it are also described.

By examining these standards it is apparent that there are significant implications for manufacturers of electrical/ electronic equipment. These implications relate to the test facilities required and the associated financial burden either in capital investment or facility hire. There are also problems associated with actually performing the tests and interpreting the standard.

6.1 EN 55 022 The Emission Standard for Information Technology Equipment

Euro-Norm 55 022[2] is the harmonised standard defining the permitted limits of radiated and conducted interference from IT equipment. The UK equivalent is BS 6527: 1988 which is dual numbered with the Euro Norm. EN 55 022[2] is based on CISPR recommendation 22[5], which is specific to IT equipment emissions. The IEC intend to adopt the test methods of CISPR 22[5] as a 'basic' standard which will be called up by future product specific emission standards. As EN 55 022[2] is invoked by the generic standards it therefore may be considered as applying to a broad range of products where the clock frequency of the electronics system is greater than 9 kHz [EN 50 081-1[3]].

In general the test methods and emission limits described in EN 55 022[2] are the same as those specified in EN 55 011[6] which applies to Industrial, Scientific and Medical (ISM) equipment. However in EN 55 011 there are also defined frequencies for unlimited radiation (this is covered in 6.4.1).

All IT equipment is a potential source of electromagnetic interference (EMI). This EMI may be unintentionally coupled into mains or signal cables or radiated into the environment. The EMI is generated by the many periodic binary pulsed waveforms which are inherent within IT equipment. EN 55 022[2] provides the manufacturer with the following information:

1. it defines what is meant by 'Information Technology Equipment '
2. it states the limits for both conducted and radiated emissions
3. and finally, it describes the methods of measurement to be used.

6.1.1 EN 55 022: Definitions

'Information Technology equipment' is defined by EN 55 022[2] as equipment used for:

- receiving data from an external source — for example, a data input line or a keyboard;
- performing processing functions on data — such as computation, transformation, recording, filing, sorting, storage or transfer;
- providing data output. This includes reproduction on a printer or screen and data sent to other equipment.

This means that virtually all electronic equipment associated with data processing functions is covered.

EN 55 022[2] goes on to define a 'test unit' as being a system comprising one or more host units. If we consider a personal computer as a system, it consists of a computer as 'host unit', together with one or more peripherals. This would be the minimum system for test purposes. A 'host unit' is defined as a unit which provides the mechanical housing for 'modules' and which may provide power distribution to other IT equipment.

'Modules' are defined as the parts of the IT equipment which provide a function. These may contain radio-frequency sources, for example, central processor unit (CPU) cards and communications interface cards, such as those using the IEEE 488 protocol.

All IT equipment is divided by EN 55 022[2] into two categories:
Class A and Class B. Paraphrasing the standard:

> '*Class A equipment* is defined as IT equipment which satisfies the Class A interference limits but which does not satisfy the Class B limits. In some countries, such equipment may be subject to restriction on both sale and use.'
>
> '*Class B equipment* is defined as IT equipment satisfying the Class B interference limits. Such equipment should not be subject to restrictions on its use.'

It is actually necessary to look at the 'Notes' to ascertain that the Class B limits are for the domestic or residential environment and that Class A is for the commercial or industrial environment. So, in effect, these definitions are identical to those adopted by the Federal Communications Commission (FCC) in the United States. However they are clearly defined by the FCC, CFR 47 part 15[7], as described in Chapter 12, but in EN 55 022[2] clarification is relegated to a footnote; creating the risk of misinterpreting the standard. The FCC is not only precise in its definitions of class A and class B but also in defining a computing or digital device as equipment having a clock rate in excess of 9 kHz. The FCC also lists equipment which is excluded from the regulations, for example transportation equipment and domestic appliances. The European washing machine manufacturer, for example, who utilises a micro-computer within the programmer is left with the unanswered question of whether it is necessary to conform to the emission requirements of EN 55 014[8] for domestic appliances only or whether it is necessary also to conform with EN 55 022[2] (EN50081 -1[3])?

The IT equipment manufacturer may wonder why, when the EMC Directive[1] is being established to remove barriers to trade, the standard refers to 'restrictions on sale and use'!

6.1.2 EN 55 022: Conducted emission limits and method of measurements

The emission limits for conducted interference are specified for both Class A and Class B equipment. They are expressed for both quasi-peak and average detector measurements. In the standard the limits are expressed in tabular form:

Table 6.1 EN 55 022: 1987 (BS 6527: 1988) Conducted Emission Limits

Frequency Range	*Limits*			
	Class A		*Class B*	
MHz	*Quasi-Peak*	*Average*	*Quasi-Peak*	*Average*
0.15 - 0 0.5	79	66	Decreasing linearly with log of frequency from 66-56	56-46
0.5 - 5	73	60	56	46
5 - 30	73	60	60	50

These limits are perhaps better expressed graphically, as in Figure 6.1.

If the average limit, as given by the standard, is met when using the quasi-peak detector, then the equipment under test automatically meets both limits and it is unnecessary to make the measurement with the average detector.

If fluctuations are observed close to the limit, then the reading must be observed for 15 seconds at each measurement frequency. The highest reading is recorded with the exception of any brief isolated reading which can be ignored.

Details of the quasi-peak and average detectors and the requirements for the receiving equipment are given in CISPR 16[9]. The measurement method is described in clause 9 of EN 55 022[2].

The equipment under test (EUT) is connected to the mains via an 'artificial mains network' this is also defined in CISPR 16[9]. This network effectively provides a stabilised mains impedance, and is known as a Line Impedance Stabilisation Network or LISN. The response of this network is such that the impedance is approximately constant from 150 kHz to 30 MHz. The measuring receiver is connected to this stabilised impedance and the measurement can be compared with the appropriate limit. Clause 9 of EN 55 022[2], details the measurement set-up as follows:

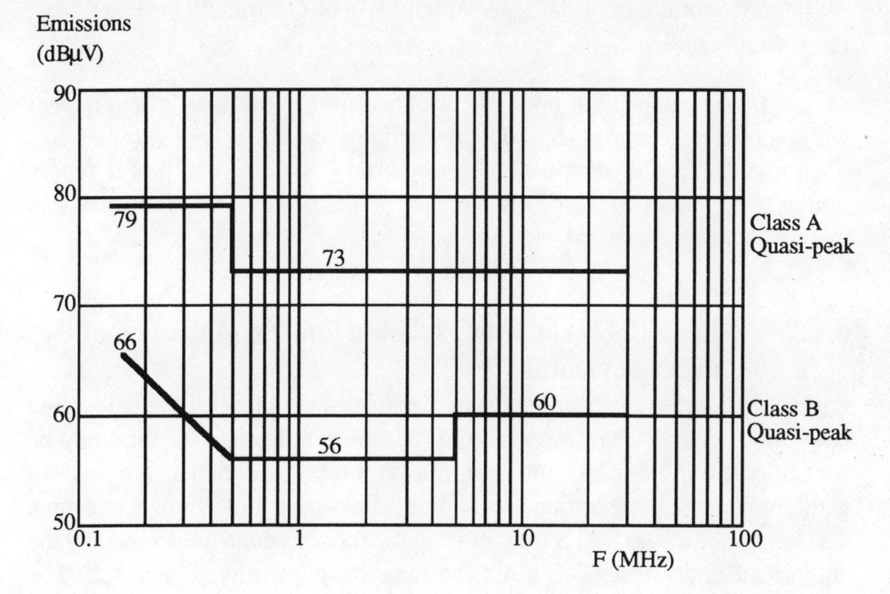

Figure 6.1 EN 55 022: 1987 (BS6527: 1988) Conducted emission limits

1) the LISN and the EUT should not be closer than 0.8 m the flexible power cord of the EUT should be 1 m long;

2) if it is in excess of this then the cable is to be folded back and forth such that the bundle does not exceed 0.4 m in length;

3) where a safety earth is required and is not provided by the manufacturer then an earth connection 1 m long should be run parallel to the mains connection and not more than 0.1 m from it. This should be connected to the reference earth of the LISN.

4) to exclude conducted ambient interference an additional filter may be connected between the LISN and the mains supply or the measurements may be performed inside a screened enclosure. In this case the screened enclosure will perform the function of a reference ground plane which is required to be a minimum of 2 m x 2 m and positioned 0.4 m away from the EUT. All other metal surfaces should be at least 0.8 m away from the EUT.

5) in the case of floor standing equipment a metal floor can be used as

the reference groundplane, this is required to extend at least 0.5 m beyond the boundary of the EUT. The EUT should also be electrically insulated from the groundplane.

6) for a system with one or more host units every item of equipment with its own power cord should be tested individually. Where several items of equipment receive their power supply from a host unit or a power supply unit, it is only the conducted interference on the power cord of the host unit that needs to be measured.

6.1.3 EN 55 022: radiated emission limits and method of measurement.

Radiated emission measurements are performed on an Open Field Test Site (OFTS) or Open Area Test Site (OATS), the implications of this may be particularly important to a manufacturer. Appropriate antennas are used together with a receiver equipped with both quasi-peak and average detectors. The boundary of the OFTS is an ellipse, the foci of which are formed by the equipment under test (EUT) and the measuring antenna (Figure 6.2) The major axis of this ellipse is twice the distance 'd' between the equipment and the antenna. The minor axis is $d\sqrt{3}$. The test site must be flat and free from overhead wires and nearby reflecting structures. The elliptical clear area defines the magnitude of reflections from outside the ellipse, such that they are small compared to the direct and ground reflected waves. The common EUT to antenna distances specified, d, are 3 m, 10 m and 30 m. EN 55 022[2] specifies limits at 10 m for class B amd 30 m for class A. The frequency range for the site is 30 MHz to 1000 MHz.

The size of the EUT limits the minimum size of site which can be used. EN 55 022[2] specifies that d must always be greater than $2D^2/\lambda$ where D is the maximum dimension of the EUT and λ is the wavelength at 30 MHz. Essentially this means that the larger the EUT the greater the EUT to antenna distance and hence the larger the site required, Figure 6.2.

The overriding factor affecting the quality of a test site is the necessity of distinguishing the emissions of the EUT from the ambient noise. The standard requires that the ambient noise is at least 6 dB below the specified limit. This requirement is relaxed when the combination of ambient noise and the emissions from the equipment under test is below the limit. At certain frequencies, occupied by say broadcast services, emissions from the EUT may be masked, as the broadcast services themselves are to be protected by the emission specification. It is ironic that they should interfere with the emission measurement.

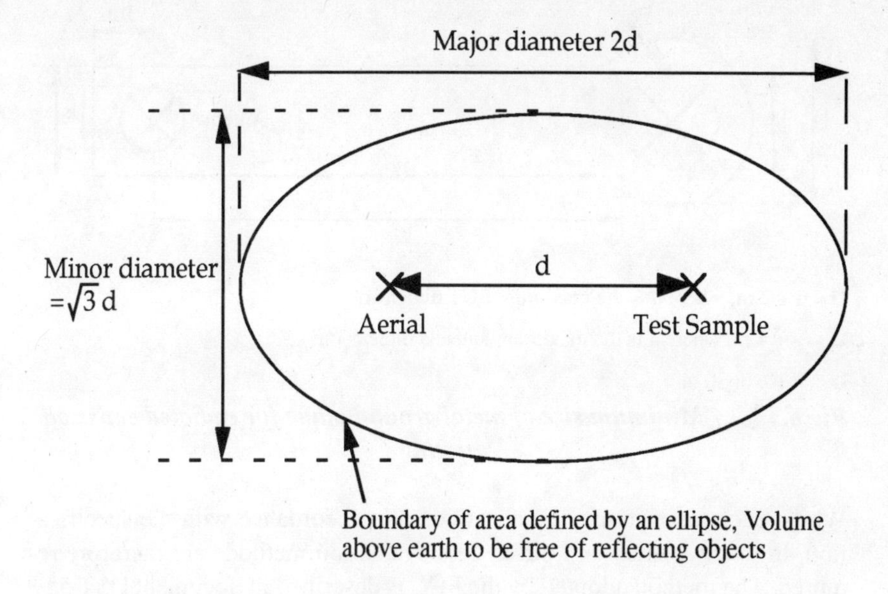

Figure 6.2 ***Radiation test site, clear area ellipse***

Because sites in different locations have differing soil and sub-soil conditions and subject to changes in climate, a ground plane is required to provide a reference to give a repeatable reflection coefficient, Figure 6.3. The standard defines its size as extending at least 1 m beyond the perimeter of the EUT and 1 m beyond the antenna; the width is defined as the maximum antenna dimension plus 1 m. In the vicinity of the EUT the width is defined as the maximum dimension of the equipment plus 2 m. The groundplane should have no voids or gaps. If mesh or perforated metal is used, then the maximum size of the perforations should not be greater than a tenth of a wavelength at 1 GHz, this is about 30 mm. If a ground plane is not used then the ground reflection is dependent mainly upon the ground conductivity which is determined by its water content. On sites without ground planes considerable performance variations can be expected due to changes in the local climate.

To find the *maximum field strength* readings the *antenna height must be adjusted between 1 and 4 m* to allow for the measurement of the direct emissions from the equipment under test and the reflected emissions from the ground plane. The *EUT* itself should also be *rotated* and the measurements should be carried out with the *antenna in both vertical and horizontal polarisations,* Figure 6.4.

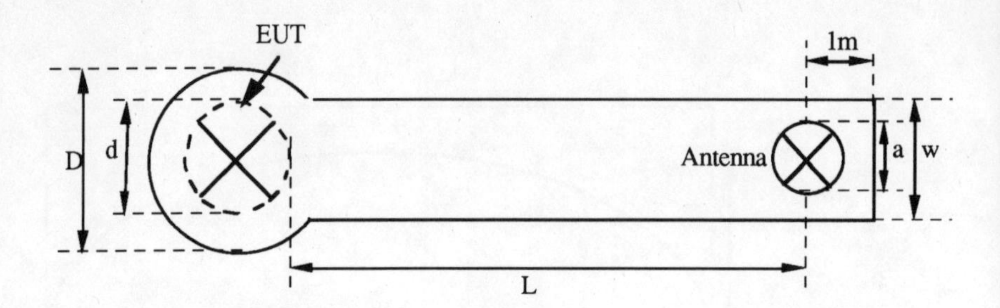

D = d + 2 m, where d is the maximum EUT dimension;

W = a + 1 m, where a is the maximum antenna dimension;

L = 3, 10, or 30 m

Fig 6.3 Minimum size of metal ground plane for radiated emission testing

Whilst each test site may be constructed in accordance with the specification, in practice sites will differ. Site evaluation methods are therefore required. The method adopted by the FCC is described in document OST 55[10] and is called 'Site Attenuation' and the VDE method is defined by VDE 0877[11]. These methods are described in Chapter 8. The test results for a given site are required to be within typically +/- 3dB of the ideal site. EN 55 022 does not define a method of measuring site quality nor does it refer to a suitable method The third edition of CISPR 16 will contain an appropriate method of validation of the radiation site and it is to be expected that this will be referred to in subsequent editions of EN 55 022. Practical implementations of open field test sites are described in Chapter 8.

The requirements for the measuring receiver, quasi-peak detector and the measurement antennas are defined in CISPR 16[9]. In practice measurements are made using both spectrum analysers and receivers. The spectrum analyser is usually much faster in operation than a receiver and is ideal for pre-compliance work, but not many analysers can meet all the CISPR 16 requirements.

Therefore for compliance testing a receiver rather than a spectrum analyser may be required. Radiated emissions are usually measured using a combination of Biconical (typically covering the frequency range 30 MHz to 200/300 MHz) and Log-periodic (typically covering the frequency range 200/300 MHz to 1 GHz) broadband antennas. Wideband antennas such as the 'BiLog' covering the whole frequency range with one antenna have been available since 1993. Tuned dipoles [CISPR16[9]] are used where more accurate measurements are required but measurements made using these are very slow.

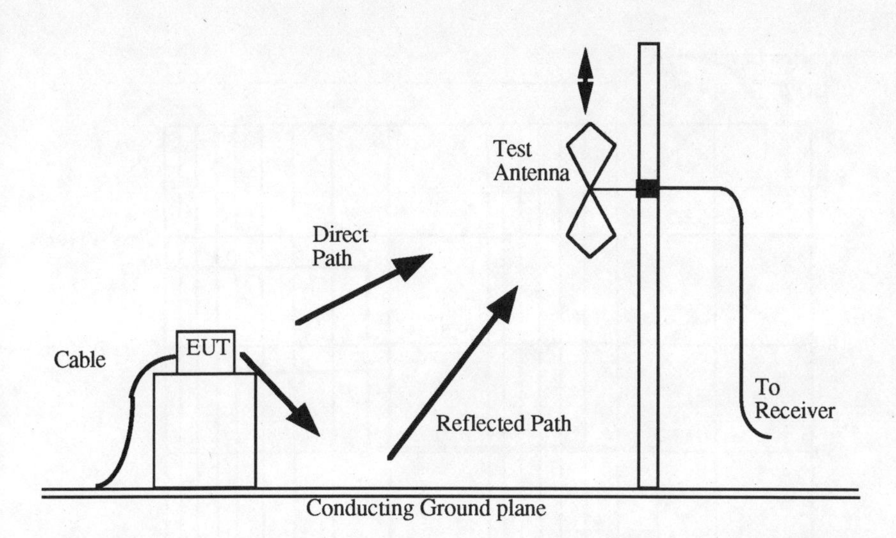

Figure 6.4 ***Open Field Test Site***

The limits for radiated emission field strengths are defined for Class A and Class B equipments in the frequency range 30 MHz to 1 GHz:

Table 6.2 ***EN 55 022: 1987 (BS6527: 1988) Radiated Field Strength Limits***

Frequency Range MHz	*Test Distance (m) Class A*	*Class B*	*Quasi-Peak Limit dBμV/m Class A*	*Class B*
30 - 200	30	10	30 (39.5)	30
230 - 1000	30	10	37 (46.5)	37

Note: Values shown in () are 10 m equivalent values for Class A.

From 30 to 230 MHz the Class A limit is 30 dBμV/m at a measuring distance of 30 m. For 230 Mhz to 1 GHz the limit is 37 dBμV/m again at 30 m. The Class B limits are defined at a measuring distance of 10 m and for the same frequency bands have the same limits. In practice it is usual to perform Class A measurements at 10 m as well. This is catered for in the standard which specifies an inverse proportionality factor of 20 dB per decade. Using this factor the Class A limits at 10 m become 39.5 and 46.5 dBμV/m respectively and are therefore 10 dB higher than the Class B limits, this is shown in Figure 6.5.

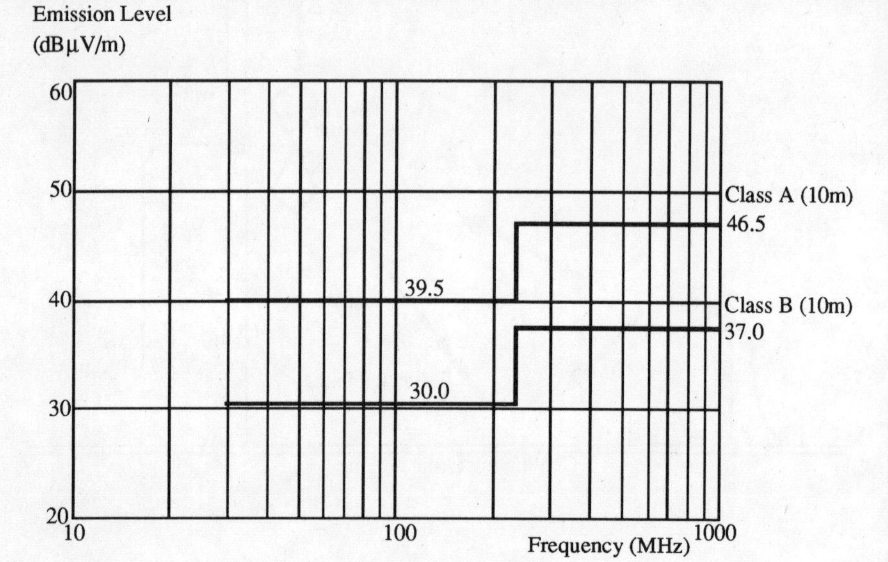

Figure 6.5 EN 55 022:1987 (BS 6527: 1988) Radiated Field Strength Limits

If the reading on the measuring receiver shows fluctuations close to the limit then, as for conducted emission measurements, the reading must be observed for 15 seconds at each measurement frequency and the highest reading recorded, ignoring any brief isolated high readings.

When performing radiated emission measurements floor standing equipment must be placed as close as possible to the groundplane. Portable equipment must be supported on a non-metallic table 0.8 m above the groundplane.

The standard accepts that it may not be possible to perform tests on sites that have all of the required features and therefore suggests an alternative. The standard states that groundplanes not satisfying all the requirements are acceptable providing that evidence can be given to show that any errors resulting from using that site do not distort the measurements.

Where it is impractical to rotate the EUT, the standard provides for it to remain in a fixed position and measurements taken by moving the antenna around it.

Should it be necessary to perform measurements on a Class A IT equipment installation, then measurements should be made at the boundary of the user's premises. If this is less than 30 m from the test unit, then the measurements must be made outside the user's premises to maintain the 30 m test distance. In practice it may be necessary to perform the measurements closer than 30 m, for example at 10 m, and to normalise the measurement to the 30 m distance.

Because the layouts of the IT equipment itself and of its associated interconnections are many and varied, the radiated emissions may also vary significantly. Because the interconnecting cables (which are physically long in comparison to the equipment itself) are likely to be the most efficient antennas, the apparently incidental lengthening or shortening of the cables can have a significant effect. EN 55 022[2] recognises and addresses this problem which makes test conditions difficult to replicate.

The EUT should be configured so as to maximise emissions, whilst at the same time remaining consistent with typical applications. Cables should be connected to the available interface ports and the effect of altering the cable positions should be investigated in order to determine the configuration which will produce maximum emissions. This configuration is to be precisely noted in the test report. The cables that are used in the testing should be the ones intended for the individual items of equipment. If the length can be varied then the length that is chosen must be the one that gives the maximum emissions.

Excess cables should be bundled such that the bundle is between 0.3 and 0.4 m in length. When this is not possible, for example because of cable stiffness or because testing is being performed at a user installation, then the cable arrangement is to be precisely recorded in the test report. All results must be accompanied by a complete description of the layout of the cables, and layout of the EUT, so that replication is possible. Any special conditions should also be noted for example cable types, grounding and shielding. These conditions must be included in the user instructions for the equipment.

Where the EUT contains a number of modules, these must be operative during the testing. The standard states that 'one of each type of Information Technology equipment that can be included in a possible system configuration shall be included in the test unit'. Where a system employs two or more identical modules or items of IT equipment it is permissible to use only one during testing. The reason for this is that, as stated by the standard: 'it has been found that emanations from identical modules of Information Technology equipment are in practice generally not additive'.

An EUT which functionally interacts with other IT equipment must be connected to it, or to an appropriate simulator. This should provide representative operating conditions. Where a simulator is used, this must represent the electrical and, if necessary, the mechanical characteristics of the interfacing IT equipment. If a 'host' has been designed to power other IT equipment then it may also be necessary to have these connected to ensure that the host will operate normally.

Interpretation of these configuration guidelines can therefore be seen as an important element in determining the emissions from an IT system or item of equipment.

6.1.4 Quantity production

For mass produced equipment it is obviously impractical and uneconomical to test each individual item. It need only be demonstrated on a statistical basis that, with at least 80% confidence, 80% of the production complies with the limits specified. This is known as the '80/80' rule. The statistical requirements are set out in section 7.2.3 of EN 55 022[2] and also in CISPR 16[9]. Alternatively a manufacturer may test a single piece of equipment and then subsequently perform similar tests on equipment taken at random from the production line. Remember that compliance with the EMC Directive is legally binding. The manufacturer may therefore need to introduce some form of production testing of a 'go/no-go' nature, or implement a quality assurance scheme to ensure consistency of production.

6.1.5 Implications

The implications for manufacturers of information technology equipment or for other manufacturers who are required to comply with EN 55 022[2] are:

• access is required to an Open Field Test Site. This implies investing in one or using a third party facility;

• although not specifically demanded, the use of a screened enclosure is implied for performing conducted emission measurements. As with an OFTS, investment in such a facility or the use of a third party facility will be required;

• access is required to suitable receivers with quasi-peak and average detectors;

• unless compliant with Class B limits, equipment may have restrictions placed on its sale or use;

•whilst the standard attempts to overcome difficulties both of interpretation and repeatability of tests, because generalisations are used in the standard, interpretations are bound to differ particularly in respect of test configuration;

• the standard makes no reference to site measurements that would ensure repeatability from one site to the next, except for the requirements on ambient noise. The FCC [OST 55[10]] regulations and VDE 0877[11], both require site attenuation measurements to be made to indicate the quality of a site. Whilst differing in detail, these regulations allow sites to be within +/- 3 dB of the ideal site attenuation. NAMAS are looking into procedures for site verification and this is also a subject being investigated by the EMC Information Technology Committee (EMCIT) of the European Organisation for Test and Certification (EOTC);

• the definitions of IT equipment 'systems' and 'installations' could be interpreted in such a way that suggests that a very large system could be required to be tested. This may be impractical except at the end users premises.

6.2 Generic Standards

CENELEC has recognised as impossible the task of providing individual standards to cover the vast range of products covered by the EMC Directive. A need has been identified for simple straightforward standards that relate to environments or locations and are not specific to products. These non-specific standards are called 'generic' standards. They are for apparatus not covered by a product or family product specific standard, which will always have precedence.

The emission standard for 'residential, commercial and light industry' locations, EN 50 081-1[3] and EN 50 081-2[4] the emission standard for 'industrial' locations, will both be considered along with the 'basic' standards which they reference.

6.3 BS EN 50 081-1; 1992 Electromagnetic Compatibility Generic Emission Standard Part 1 Residential, Commercial and Light Industry

This generic standard will cover a large proportion of the products falling within the scope of the EMC Directive. The generally recognised breakpoint between 'residential', 'commercial' or 'light industrial' and 'industrial' is whether or not a device is connected to a low voltage 'mains'.

The environments encompassed by this standard are: residential, commercial and light-industrial locations, and apply to both indoor and outdoor sites. Examples given in the standard include:

- residential properties, such as houses and apartments
- retail outlets: shops and supermarkets
- light-industrial locations: laboratories, workshops and service centres
- business or commercial premises, such as offices and banks
- outdoor locations: car parks, petrol stations, amusement and sports centres
- areas of public entertainment: cinemas, public bars and dance halls

The installed equipment is assumed to be directly connected to low voltage mains supplies.

The housing and external connections of equipment which interface with the outside electromagnetic environment are defined as ports. The five ports defined by the standard are:

- the enclosure port
- the ac power port
- the dc power port
- earth connections
- and the signal, control or other connections

These ports are illustrated in Figure 6.6.

This emission standard may be applied to equipment or apparatus which may be used within the 'residential, commercial and light-industry' location. Reference is made to published international standards for methods of measurement. These methods may need to be adapted to cater for particular product groups. The tests are defined in tables and are applicable to the ports defined in Figure 6.6.

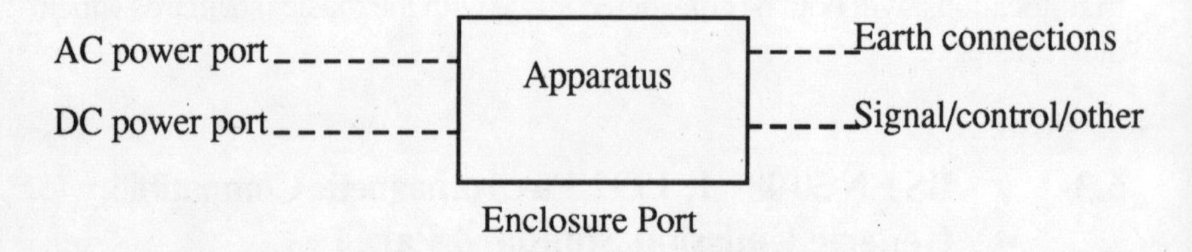

Figure 6..6 Apparatus port definition for EN 50 081-1

The emission limits and measurement methods are as defined for EN 55 022[2], Class B. In addition, reference is made to EN 55 014[8] for relaxed limits in respect of discontinuous or infrequently repeated noise, and to EN 60 555 parts 2[12] and 3[13] for harmonic content and voltage fluctuations.

Further applicability guidance is given:

- measurements are made on the relevant ports of the apparatus according to Table 1 of EN 50 081-1, (see Table 6.3)
- measurements are only required where the EUT has the relevant ports
- radiated emission measurements are required only for apparatus con taining processing devices operating at frequencies higher than 9 kHz
- harmonic and voltage fluctuation measurements are only applica ble to equipment within the scope of EN 60 555 Parts 2 and 3
- it may be determined from the electrical characteristics and the

usage of a particular apparatus that *some measurements are inappropriate and therefore unnecessary.* In such cases the decision not to measure must be recorded in the test report.

Table 6.3 Emission limits specified by EN 50 081-1

	Frequency range	Limit (Ref. Std)	Port
Harmonics	0 - 2 kHz	EN 60 555-2, -3	AC power
RF Voltage	150 kHz - 30 MHz	EN 55 022/B	AC power
RF Voltage	Discontinuous	EN 55 014	AC power
Field Strength	30 MHz -1 GHz	EN 55 022/B	Enclosure

The measurements should be made in the operating mode producing the largest emission within the frequency band being investigated. Other configuration requirements are similar to those described for EN 55 022, excepting that a 'generalised' EUT is being considered not ITE.

The standard also specifies documentation which must be supplied to the purchaser/user, this must contain information on any special measures taken to achieve compliance, *eg*., the use of shielded cables. Also a manufacturer on request, should be in a position to provide a list of auxiliary equipment which together with the apparatus comply with the emission requirements.

An informative annex is included which suggests that conducted interference may be measured on signal, control, and dc power lines using a current probe with the appropriate line terminated in common mode to the ground plane *via* 150 ohm. The reference document specified for this is an October 1989 amendment to CISPR publcation 22[14]. The limits specified are the class B limits. For 150 to 500 kHz the limit decreases linearly with the log of frequency, 40-30 dBμA quasi-peak and 30-20 dBμA average. For the frequency range 500 kHz to 30 MHz the limits are 30 dBμA and 20 dBμA, quasi-peak and average respectively.

Whilst this annex is 'informative' in the first edition of EN 50 081-1, it is likely to be adopted as a requirement for the next issue.

Three basic or reference standards are called up by EN 50 081-1. These are EN 55 022[2], EN 60 555 (parts 2[12] and 3[13]) and EN 55 014[8], as has been already indicated. EN 55 022 has been described in some detail in section 6.2 of this chapter. An overview is now provided for EN 60 555 and EN 55 014 in relation to the generic standard.

6.3.1 EN 60 555 'Disturbances in supply systems caused by household appliances and similar electrical equipment', Part 2: harmonics

For the frequency range 0-2 kHz EN 60 555 part 2[12] is called up by the generic standard EN 50 081-1[3] to define the maximum permissible values of harmonic components of the input current produced by the EUT under *specified conditions* (principally the supply requirements for the tests).

The harmonics of the input current are measured in accordance with clause 5 [EN 60 555-2[12]] and should not exceed the values calculated from [Table 1], reproduced as Table 6.4. These values apply to line and neutral currents for a 230 V line-neutral or 400 V line-line supply operating at 50 Hz. The limits may be applied to other system voltages when multiplied by the system voltage and divided by either 230 or 400 respectively. When applied to portable tools the limits are multiplied by 1.5.

The limits apply to either transitory or steady state conditions, with the exception of non-repetitive harmonic currents generated for a few seconds during equipment start-up. Limits are also under consideration for: repetitive transitory harmonics caused by switching equipment motors; television receivers; and battery chargers for electric road vehicles.

Measurements are carried out by connecting the EUT to a supply having the same nominal voltage and frequency rating as the EUT. The current in each line (and neutral 3ϕ) is monitored by either a non-inductive shunt (having a resistance and time constant of less than 0.1 ohm and 10^5 sec respectively) or a current transformer with a negligible series resistance (and calibrated for each harmonic frequency). If the harmonic components are found to vary more than proportionally with the supply voltage Vs, then tests at 0.94 Vs and 1.06 Vs are to be carried out. For an EUT with more than one rated voltage the test should be carried out at the higher voltage *ie*. the 'voltage producing the highest harmonics as compared with the limits'.
During testing equipment should be operating under normal load and normal operating conditions.

The power supply source requirements are demanding, such that in practice, they are only likely to be met by a 50 Hz power amplifier. During measurements the voltage is to be maintained within +/- 2% and the frequency within +/- 0.5% of nominal. The internal impedance should be sufficiently low so that it does not interfere with the harmonic measurement. Ideally this would be zero impedance. However in practice, as long as the internal impedance does not affect the harmonic measurement by more than 5% of the permissible limit, it is acceptable.

Table 6.4 Limits of harmonics

Harmonic Order	***Max allowed harmonic current (A)***
Odd Harmonics	
3	2.30
5	1.144
7	0.77
9	0.4
11	0.33
13	0.21
15<n<39	$0.15 \times \frac{15}{n}$
Even Harmonics	
2	1.08
4	0.43
6	0.3
8<n<40	$0.23 \times \frac{8}{n}$

From appendix A of [EN 60 555-2[12]] the source reference impedance that has been used is:

Zs = 0.4 + j*n* 0.25 ohms (where *n* is the order of the harmonic)

Care is also required to ensure that resonances of the source inductance with the EUT capacitance does not occur. A further requirement is that the harmonic ratio of the voltage supplied by the source at no load (V_{SNL}) to the on load voltage (when supplying a resistive load corresponding to the EUT rating, V_{SRL}) should be small. The guideline given for the 1/3rd harmonic for example is:

$V_{SRL}/V_{SNL} \times 100 = 0.9\%$
for the 1/5th harmonic it is 0.4%, *etc.*

EN 60 555-2 also specifies the requirements of the harmonic analyser which may be used. Various instruments are suggested including a spectrum analyser. The measurement error permitted is 5% of the specified limit for all harmonic current components between 2 and 40. The selectivity of the instrument is defined for each harmonic frequency: for example for an injected frequency of 250 Hz, the attenuation for frequencies below 200 Hz and above 300 Hz should be greater than 30 dB. Also the response is defined for a 1 second application of a sinusoidal voltage at each harmonic frequency and the 'transient' response to a suddenly applied sinusoidal voltage.

6.3.2 EN 60 555 'Disturbances in supply systems caused by household appliances and similar electrical equipment', Part 3: Voltage fluctuations

This standard defines limits for the voltage fluctuations impressed on the public low voltage supply system and is called up as a reference standard by the generic standard for the 'residential, commercial and light industry environment' [EN 50 081-1[3]]. The limits defined are based on the subjective 'flicker' observed from a 230 V/60 W coiled-coil filament lamp during a 10 minute observation period.

A 'flickermeter' has been developed because flicker is subjective and a standard measurement is required. This instrument is designed to measure quantities related to luminance fluctuation. It simulates the process of physiological visual perception and gives a reliable indication of the reaction of an observer to any type of flicker. Functional and design specifications are described in IEC 868 [IEC 868, 1986[15]].

EN 60 555-3 defines four types of voltage fluctuation :

i) periodic rectangular voltage changes
ii) step changes of voltage which are irregular with time (less than 1000/minute)
iii) clearly separated voltage changes which are not all step changes *eg*. switching non-resistive loads
iv) a series of random or continuous voltage fluctuations *eg*. 10 Hz amplitude modulation

The limits are defined by Figure 4a of EN 60 555-3[13] in terms of the maximum permissible percentage voltage change, $\Delta U/U\%$ as a function of voltage changes per minute (or/second). These voltage changes range from for example 3% for voltage changes occuring between 0.1 and 0.76 per minute to 0.29% for voltage changes occuring 1052 times per minute.

The testing itself should be performed for the EUT operation which produces the most unfavourable voltage fluctuation. The EUT is connected to an ac supply having a defined source impedance and the flickermeter is connected between line and neutral. As for EN 60 555-2[12] the single phase supply source impedance is specified to be $0.4 + jn0.25$ ohms.

A draft revision of IEC 555-3 was issued for public circulation and comment in August 1990 [90/28296[16]] and a Technical Report was issued by the IEC in April 1991 [IEC 868-0: 1991[17]]. Although pre-dating the technical report, the draft revised standard incorporates the flicker evaluation procedures described in the report. This method defines the short-term flicker severity 'P_{st}' and the long-term flicker severity'P_{lt}' for observation periods of 10 minutes and 2 hours respectively. The new limits proposed are based on flicker severity being evaluated over a short period with P_{st} =

1 being defined as the conventional threshold of irritability. P_{lt} is based on successive values of P_{st} and the proposed limit is 0.65 [90/28296[16]]. The draft revised standard defines the supply impedance and in Appendix A [90/28296[16]] details test conditions for various items of domestic apparatus: cookers, hotplates, baking ovens, grills, oven/grill combinations, microwave ovens, lighting equipment, washing machines, clothes dryers, refrigerators, copying machines/laser printers, vacuum cleaners, food mixers, portable tools, hairdryers, 'brown goods' and direct water heaters.

The report has taken a statistical approach to take account of the mechanisms of vision and the building up of annoyance by evaluating the effects over a sufficiently representative time period. Some loads cause flicker of a random nature and the instantaneous level can be wildly and unpredictably variable. Therefore it is not just the maximum attained levels of flicker that are important and consideration has been given to what fraction of the observation period any given flicker level was exceeded. The report includes the 'changes' permitted in IEC 555-3 [EN 60 555-3[13]] and compares them against changes producing one unit of flicker severity ($P_{st} = 1$) for various changes of voltage per minute using the 'five point algorithm' being evaluated. The report determines five percentile levels required for evaluating flicker severity and applies this 'classifier' to the flickermeter. It defines six calibration points of $\Delta U/U\%$ for the flickermeter, when a rectangular wave, having between 1 and 1620 changes/minute is applied to it.

For equipment exceeding 16 A rating or subject to special consent from a supply authority, recommendation for voltage fluctuations are included in the draft IEC 555-5 [90/28293[18]]. This is again based on the P_{st} and P_{lt} measurements.

It should be noted that, in addition to the limits and how they are applied, the scope of IEC 555 (and hence EN 60 555) is under consideration and may well be wider ranging in the future.

6.3.3 EN 55 014 'Limits and methods of measurement of radio interference characteristics of household electrical appliances, portable tools and similar electrical apparatus'

The generic standard for the 'residential, commercial and light industry' environment [EN 50 081-1[3]] innocuously calls up EN 55 014: 1987[8] as the reference standard for defining and determining conducted 'discontinuous' interference in the frequency range 0.15 to 30 MHz for the ac power port. When the reader actually picks up EN 55 014 (or BS 800 by which it was more familiarly known in the UK) he or she is confronted with a 60 page document written in 'standardese'!

It is not intended that a complete guide will be given here to the limits for discontinuous interference or to the methods of measurement, but an attempt will be made to provide the reader with some of the key pointers to deciding whether discontinuous interference limits apply to his/her equipment and how to use the standard.

Switching operations in thermostatically controlled appliances, automatic programme controlled machines and other electrically controlled or operated equipment may generate discontinuous interference. The subjective effect of discontinuous interference depends on the the repetition rate and amplitude for both radio and television receivers or 'Hi-Fi' equipment. Therefore distinctions have to be made between the various kinds of discontinuous interference.

It is important to refer the reader to Appendix D of EN 55 014[8] which provides 'guidance notes for the measurement of discontinuous interference' and also to 'Figure. 12 - Flow diagram for measurement of discontinuous interference (see Appendix D)' of EN 55 014[8] reproduced here as Figure 6.7. However before sense can be made of these a number of definitions need to be mastered.

- ***Discontinuous interference*** — electromagnetic interference occuring during certain time intervals separated by interference-free intervals [161-02-13, 1990[19]].

- ***Click*** — an electromagnetic disturbance which, when measured in a specified way, has a duration not exceeding a specified value [161-02-15, 1990[19]].

For EN 55 014 the click duration is specified as being less than 200 ms and the interference free interval as greater than 200 ms.

- ***Counted clicks*** (n_1) — clicks which exceed the limit of continuous interference.

The limits of continuous interference are almost identical to EN 55 022 (Table 6.1 and Figure 6.1); the quasi-peak limits are as for ITE, whilst the average limit decreases linearly with frequency from 59 dBμV (not 56) to 46 dBμV and higher limits are allowed for portable tools in three ratings up to 2 kW.

- ***Click rate*** (N) — the click rate determines the discontinuous interference limit and is the number of clicks per minute, this should be determined for the frequencies:

160 kHz, 550 kHz, 1400 kHz and 30 MHz:

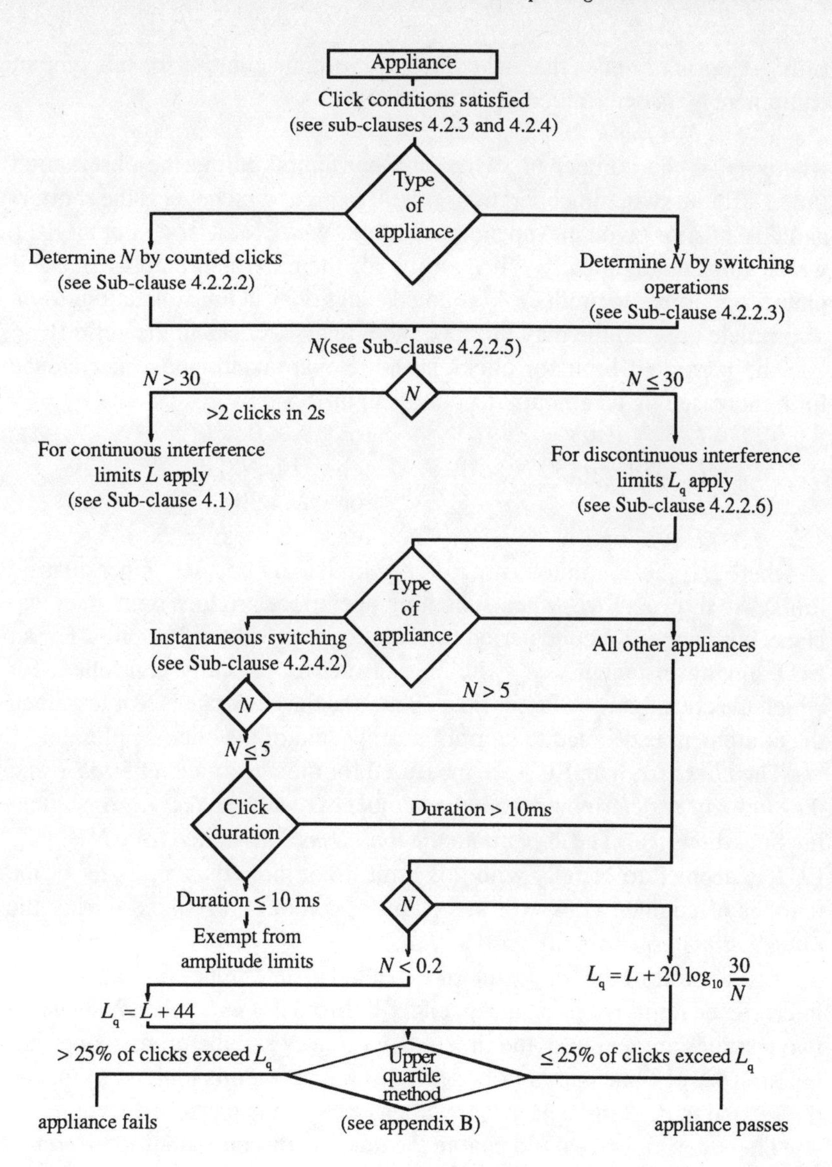

Figure 6.7 Flow diagram for measurement of discontinuous interference

$N = n_1/T$

— *where*:T is the *observation time* in minutes and is defined as the time to register 40 counted clicks or is 120 minutes whichever is the shorter. Some apparatus functions by switching on and off either, for example, under thermostatic control (refrigerators) or under manual control (sewing machines), for this type of equipment N is determined as follows:

$$N = fn_2/T$$

where n_2 is the number of *switching operations* during the observation time T (for 40 switching operations or 120 minutes whichever is the shorter) and f is a factor given in Appendix A, Table IV of EN 55 014[8]. For an EUT which stops automatically T is the time for the minimum number of complete programmes to produce 40 counted clicks or switching operations (*note* a complete programme may produce more than 40 clicks in the period).

The permitted limit for clicks is the relevant continuous interference limit increased by an amount dependent upon N:

$L_q = L + 44$ dBμV for $N < 0.2$
$L_q = L + 20\log_{10}(30/N)$ dBμV for $0.2 < N < 30$
$L_q = L$ for $N > 30$
$Lq = L$ for >2 clicks in 2s

— where L is the continuous interference limit and Lq is the 'upper quartile limit'. As shown above, when switching operations produce more than two clicks within any 2 second period then the continuous limit L applies. For an EUT having instantaneous switching *eg*. thermostatically controlled, for which the click duration is less than 10 ms and the click rate is 5 or less, then the equipment is deemed to comply independent of the click amplitude.

The *clicks* from an EUT are measured for the *observation period T* and the *click rate N* determined from the number of *counted clicks n_1* (or switching operations n_2). The *upper quartile limit L_q* is determined from N and the EUT is deemed to comply with this limit if not more than a quarter of the number of counted clicks (or switching operations) registered during the *observation period* are higher than L_q.

An example of the 'upper quartile method to determine compliance with interference limits' is given in Appendix B. From this example it is seen that two test runs are required, the first to determine N and therefore the permitted limit L_q, also the number of clicks allowed above this limit *ie*. $n_1/4$, and the second to determine how many clicks exceed the permitted level.

These definitions should enable the guide to discontinuous interference [Appendix D, of EN 55 014[8]] and the flow diagram Figure 6.7 to be followed.

Measurements are performed at frequencies of 160 kHz, 550 kHz, 1400 kHz and 30 MHz using measuring instruments specified in CISPR 16[9]. The receiver is specified in section 1, the artificial mains network (LISN) in section 2, these are used in conjunction with a storage oscilloscope or the disturbance analyser for automatic measurement and assessment of the measuring results described in clause 30 and Appendix R of CISPR 16[9]. The precautions specified when using the LISN are similar to those described for the ITE conducted interference mesurements (6.1.2). Externally generated disturbances not produced by the EUT should measure at least 20 dB below the lowest voltage limit it is required to measure *ie.* as for ITE measurements, an additional filter may need to be connected in series with the LISN.

During testing an EUT should be operated under normal load conditions. These are defined for a wide range of household appliances and portable equipment in the twelve pages of section 5.3, EN 55 014[8]. These loading conditions may be helpful to manufacturers of other equipment requiring to comply with the generic standard. For hand held equipment without an earth it may be necessary to use an 'artificial' hand of a wrapping of metal foil connected *via* a 220 pF capacitor and 510 ohm resistor to the reference ground. This is illustrated in Figure 7 of EN 55 014[8].

6.4 EN 50 081-2 'Electromagnetic Compatibility — Generic Emission Standard Generic Class: Industrial'

CENELEC has adopted a further generic emission standard for the 'industrial' environment, EN 50 081 Part 2[4].

This follows the same format as for part 1 of the standard but has less onerous emission limits. Reference is again made to published standards, principally EN 55 011[6].

The informative annex for the measurement of conducted emissions on signal, control and dc power lines is the same as for EN 50 081-1[3] (see 6.3), invoking the CISPR 22 October 1989 amendment[14] excepting that the class A limit is specified *ie.* for the frequency range 150 kHz to 500 kHz the quasi-peak and average limits are respectively 53 to 43 and 40 to 30 dBμA decreasing with the log of frequency and for 500 kHz to 30 MHz 43 and 30 dBμA respectively.

6.4.1 EN 55 011 'Limits and methods of measurement of radio disturbance characteristics of industrial, scientific and medical (ISM) radio-frequency equipment'

EN 55 011: 1991[6] defines the 'Limits and methods of measurement of radio disturbance characteristics of industrial, scientific and medical (ISM) radio-frequency equipment'. ISM equipment or appliances are designed to generate or use locally radio frequency (RF) energy. EN 55 011 is a product specific standard for ISM equipment but it is referenced by the generic emission standard for the *'industrial'* environment [EN50081-2[4]] as a basic standard and therefore will apply to a large range of apparatus which is used within this environment.

EN 55 011 has much in common with EN 55 022[2], the product specific emission standard for ITE which is used as a basic standard by the generic emission standard for the 'residential, commercial and light industry' environment and has already been considered in detail within this chapter. Three aspects of EN 55 011 will be considered:

1. a comparison between EN 55 011 and EN 55 022 identifying the common areas
2. those aspects of EN 55 011 which will affect likely users of the generic standard EN 50 081-2
3. considerations for ISM equipment manufacturers.

6.4.1.1 Comparison of EN 55 011 and EN 55 022 identifying common areas

Equipment classification — although the precise wording in the two standards of the definitions of Class A and Class B equipment differs, in essence they are the same, *viz*. Class A equipment is for use in a non-domestic environment (*eg* industrial). Class B is for equipment used in a domestic environment or which is directly connected to a low voltage power supply network which can be used for domestic purposes (*ie* if equipment can operate off a 13 A socket outlet it is Class B). It is interesting to note that, whilst ostensibly this is an equipment classification, it is in reality an environmental classification.

In general the methods of emission measurements described in EN 55 011[6] are identical to those used by EN 55 022[2] already described in section 6.1 of this chapter. The emission limits defined are also identical in some areas but the limits for EN 55 011 are more extensive recognising that ISM equipment by virtue of its use of RF energy, has the inherent capability of generating large EM disturbances.

Conducted emission limits — the conducted interference limits for Class B equipment are identical in both standards. The EN 55 022 Class A limit and the EN 55 011 ClassA 'Group1' (*see* section 6.4.1.3 for definition of 'Groups') limits are also identical. Some types of ISM equipment are inherently more likely to produce higher levels of EM disturbance and are defined as Group 2 in consequence. EN 55 011 defines a further set of limits approximately 20dBμV higher than Group 1 [Table IIA EN 55 011[6]]. EN 55 011 states that limits are also 'under consideration' for the frequency range 9 kHz to 150 kHz.

The measurement method for the two standards is identical ie an 'artificial mains network' (LISN) is used along with a suitable receiver with both quasi-peak and average detectors, as specified in CISPR 16[9]. EN 55 011 also allows the use of a 'voltage probe' for conducted emission measurements (*see* section 6.4.1.2 and Figure 6.8).

The measurement set-up specified in EN 55 011 [clause 7.4.2] is similar to that specified in EN 55 022 [clause 9] and described in section 6.1.2 of this chapter.

Radiated emission limits — the EN 55 022 Class A and B limits for the frequency range 30 MHz to 1 GHz are identical to the EN 55 011 Group 1 Class A and B limits. In addition EN 55 011 defines:

- the Group 1 Class A limits for *in situ* measurements (*see* section 6.4.1.2),
- higher limits at certain frequencies for Group 2 Class A and B equipment,
- limits to 'protect specific safety services',
- also limits are under consideration for the frequency ranges 9 kHz - 150 kHz, 150 kHz-30 MHz, 1 GHz-18 GHz {excepting 11.7-12.7 GHz where the limit is defined as 57 dB(pw) erpreferred to a half-wave dipole} and 18 GHz-400 GHz.

The method of measurement defined by EN 55 011 is identical to EN 55 022 (see 6.1.2), *ie* the use of an open field test site (Figure 6.2), suitable antennas and a receiver having quasi-peak and average detectors. The receiver, detectors and antennas are as defined in CISPR 16[9].

Conformity of production — both EN 55 011 and EN 55 022 permit conformity of production to be demonstrated by the 80/80 rule (see 6.1.4). However EN 55 011 specifies that 'equipment not produced in series shall be tested on an individual basis' and further that 'each individual equipment is required to meet the limits....'.

Configuration of the EUT and load conditions — the provisions within the two standards are in essence the same. The test configuration should be consistent with typical applications whilst at the same time ensuring that the maximum emissions are measured. Both standards stress the need to include a complete description of the cable and equipment orientation within a test report so that tests may be replicated. The EUT is required to operate under load conditions, EN 55 011 [clause 7.5[6]] describes in detail typical test loads which might be employed for various types of ISM equipment.

6.4.1.2 EN 55 011 and EN 50 081-2, in particular in situ *testing*

EN 50 081-2[4] references EN 55 011[6] as a basic standard. The limits defined by EN 50 081-2 are as for Group 1 Class A for the frequency ranges 150 kHz to 30 MHz conducted emissions and 30 MHz to 1 GHz radiated emissions (these are also the same as for EN 55 022[2] Class A). The radiated limits are defined for an antenna to EUT separation of 30 m, however EN 50 081-2 permits measurements at 10 m with an increase in the limits by 10 dB they are therefore as shown for the Class A limit in Figure 6.5.

As discussed in Chapter 5, there is a lack of product specific standards for large equipment or systems and, as most of these are used within the industrial environment, manufacturers will look to use the generic industrial standard. EN 50 081-2 makes no specific provision for large equipments other than by inference but does not permit *in situ* testing:

1) 'Apparatus installed in the locations covered by this standard is considered to be connected to an industrial power distribution network and not connected to low voltage public mains supplies.'
2) the definition of ports is the same as for EN 50 081-1[3], Figure 6.6, with the addition of the process measurement and control port, therefore because of its generic nature this definition does not exclude large equipments
3) the description of locations includes the following conditions:
   - a part of the load is fed through converters
   - ISM apparatus *eg* welding machines is present
   - heavy inductive or capacitive loads are frequently switched
   - loads are rapidly varying
   - the EUT is intended to operate at less than 1000 V ac

Since EN 55 011[6] is called up to provide the methods of measurement of both conducted and radiated emissions, it is reasonable that the generic industrial standard should reference the methods for *in situ* measurements for large apparatus. Unfortunately this is not the case and for large installed equipment it will be necessary to use a TCF to demonstate conformity. In clause 5 of EN 55 011[6] there is a note which states that 'due to size, complexity or operating conditions some ISM equipment may have to be measured *in situ* in order to show compliance...', it may therefore be considered reasonable to apply the *in-situ* mesurement methods and limits to other large equipment and include the results within a TCF.

Conducted emissions — in clause 5, EN 55 011[6], it is stated that an *in situ* conducted limit for Class A equipment is 'under consideration'. This is not very helpful, however provision is made for measuring conducted interference using a 'voltage probe' when the LISN cannot be used [clause 7.2.3[6]]. The probe is shown in Figure 6.8 and consists of a blocking capacitor and a resistor such that the total resistance between line and earth is at least 1500 ohms. The probe is used by being connected sequentially between each line and the chosen reference earth.

Radiated emissions — low frequency emission limits, 9 kHz to 150 kHz and 150 kHz to 30 MHz, are under consideration and are to be measured using a loop as defined in CISPR 16[9]. This may be important for large installations; for example, strong magnetic fields can be detected in the vicinity of power electronics equipment or ac mains supply cables carrying large currents.

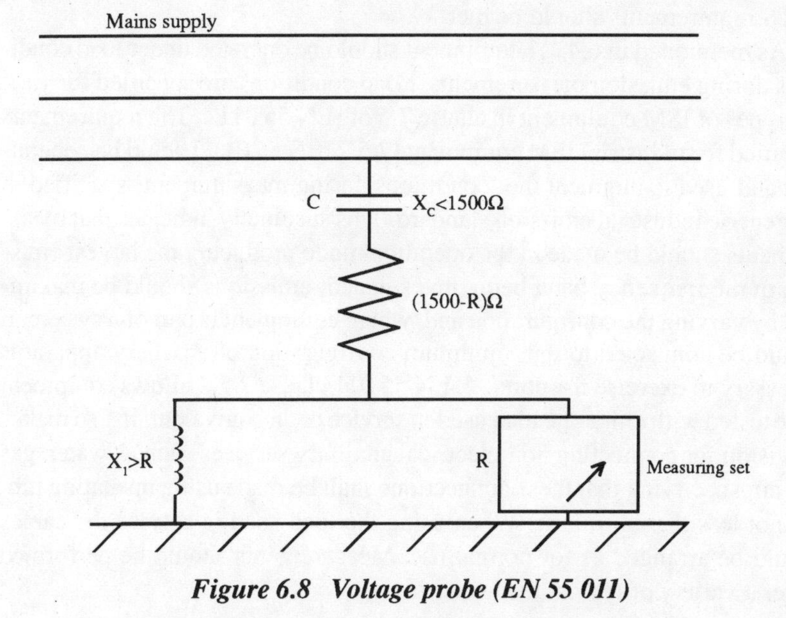

Figure 6.8 Voltage probe (EN 55 011)

For equipment not measured on an OFTS, measurements should be made after the equipment has been installed at the user's premises [EN 55 011[6], clause 10]. Measurements are made from the exterior wall outside the building housing the equipment at 30 m, or 10 m for some frequencies specified in EN 55 011 [Table VI[6]]. For the frequency range *30 MHz to 1 GHz in situ limits* are specified for Group 1 Class A [EN 55 011, Table III[6]], defined for a *measuring distance of 30 m from the exterior wall* of a building in which the equipment is situated. These limits are the same as for Group 1 Class A [Table III[6]] and Class A ITE [EN 55 022[2]] (see Table 6.2 and Figure 6.5) measured on an OFTS, *ie.* for *30 - 230 MHz, 30 dBμV/m and for 230 - 1000 MHz, 37 dBμV/m.* The measurements should be made in as many positions 'as reasonably practical but there shall be at least four measurements in orthogonal directions, and measurements in the direction of any existing radio systems which may be adversely affected'. These provisions for *in situ* measurements can be applied to equipment used in the 'industrial' environment and results included in a TCF.

Clause 8.1.3 of EN 55 011[6], whilst not specifically referring to large equipments, allows for the antenna to be moved around the EUT when the

EUT is not rotated on a turntable. At each position measurements are to be made with the antenna in both horizontal and vertical orientations. The directions of maximum radiation should be identified and the highest level at each frequency recorded. At each position of the measurement antenna the OFTS requirements should be met.

As mentioned in 6.4.1.1 equipment should be operated under load conditions during emission measurements. Load conditions are specified for various types of ISM equipment in clause 7.5 of [EN 55 011[6]]. The requirements specified for industrial ISM equipment [7.5.2, EN 55 011[6]] could be generalised and used to augment the 'conditions during measurement' specified in the generic industrial emission standard. This succinctly indicates that measurements should be made in the operation mode producing the largest emission in the frequency band being investigated; emissions should be maximised by varying the configuration and, where equipment is part of a system, it should be connected to the 'minimum configuration of auxiliary apparatus necessary to exercise the ports...'. EN 55 011 clause 7.5.2 allows equipment to be tested with either the load used in service or an equivalent. It also makes provision for connecting non-electrical auxiliary services such as water, gas and air, specifying that these connections shall be made using insulating tubing not less than 3 m long. When using the load used in service the cables should be arranged as for normal use. Measurements should be performed under a variety of typical load conditions.

6.4.1.3 Considerations for ISM equipment manufacturers

ISM equipment is equipment intentionally using RF energy for the necessary functioning of the equipment itself. As such it has the potential for causing EMI and therefore there are strict controls on the fundamental frequencies which can be utilised by this equipment. The International Telecommunications Union (ITU) has allocated specific frequencies for ISM use. These are listed in EN 55 011[6] Table Ia. Although some of these frequencies are listed as being 'unrestricted', reference to Table Ib shows that in some European countries this is not the case, *eg*. the frequency 13.56 MHz is defined as unrestricted by the ITU but is restricted to a field strength limit of 110 dBμV/m at 100 m in the UK. So ISM equipment manufacturers should take care when selecting the RF operating frequency of their equipment.

Before being able to assess the emissions from a particular item of ISM equipment, the manufacturer must firstly define to which grouping of equipment it belongs. ISM equipment is separated into two groups:

> ***Group 1***: ISM equipment in which intentionally generated RF energy is conductively coupled. General headings for equipment in this group are Laboratory, Medical and Scientific, examples are: signal generators,

measuring receivers, frequency counters, flow meters, spectrum analysers, weighing machines, electron microscopes, switched mode power supplies (when not incorporated in an equipment).
Group 2: ISM equipment in which the intentionally generated RF energy is used in the form of EM radiation for the treatment of materials. General headings for equipment in this group are Industrial induction heating equipment, Domestic induction cookers, Dielectric heating equipment, Industrial microwave heating equipment, Domestic microwave ovens, Medical apparatus, RF excited welding apparatus, Spark erosion equipment, Thyristor-controlled equipment and spot welders. Examples of processes covered: metal melting, component heating, soldering and brazing, tube welding, woodglueing, plastic pre-heating, food processing, biscuit baking, food thawing, paper drying, textile treatment, adhesive curing, material pre-heating, short-wave therapy equipment, microwave therapy equipment [EN 55 011[6], clause 4.1 and Annex A].

These are *examples* of equipment types and processes *only* and are not intended to be exhaustive or restrictive. It can be seen from these two groups of equipment that Group 2 equipment is inherently more likely to generate greater emissions, particularly radiated emissions.

Having identified the grouping which the equipment belongs to, the Class must be defined, *ie* Class A 'industrial' or Class B 'domestic' (*see* 6.4.1.1). The emission limits can then be determined from the various tables.

For Group 1 Class A and B equipment the limits of radiated and conducted emissions are defined by EN 55 011 in Tables IIA, IIB and III (6.4.1.1). The limits for Group 2 Class A and B are defined by EN 55 011 Tables IIA, IV and V. Tables IV and V are more complicated and specify particular limits for particular frequency ranges. If the radiated emission limits for Group 1 Class A [Table III] are compared with the Group 2 Class A [Table V] limits, between 30 and 230 MHz the Group 2 limits are between 10 and 38 dBμV/m higher and between 230 MHz and 1 GHz, the Group 2 limits are between 13 and 16 dBμV/m higher. Table V also specifies *in situ* limits for a measuring distance of 30 m from an exterior wall outside the building housing the equipment, these allow for attenuation by the building of 10 dB, *ie*. the limits are 10 dBμV/m lower than those for use on an open field test site.

A further constraint is placed on the limits for radiated emissions by Table VI. This defines the limits required 'to protect specific safety services in particular areas' and are defined for *in situ* measurement. *These limits may be required by national authorities.*

The effect of Table VI on the limits laid down for Group 2 Class A equipment in Table V, can be illustrated by example, this is shown in Table 6.5.

From Table 6.5 it is apparent that for the three examples chosen, the frequency ranges do not coincide but overlap and that, whilst for an emission of 500 kHz the limits are identical, for an emission of 100 MHz the safety service limit is 20 dB tougher than the Group 2 Class A limit, taking account of the measuring distance. For an emission at 1 GHz the safety service limit is 13 dB tougher than the Group 2 Class A limit.

The safety service limit in fact coincides with the Group 1 and ITE Class A limits for the frequency range 30 MHz to 1 GHz when the 10 dB attenuation factor of a building is taken into account. Therefore in Group 2 Class A applications, where it is thought that the national authorities might insist on the safety service limit, manufacturers or the end user should ascertain this at the outset of a project otherwise compliance with these limits may be difficult to achieve.

Table 6.5 Comparison of Group 2 Class A ISM limits and specific safety service limits

Frequency range MHz	Group 2 Class A Limit dBμV/m *in situ* @ 30m	Frequency range MHz	Table VI Limit dBμV/m *in situ* @ 30m
0.49 - 1.705	65	0.2835 - 0.5265	65
87 - 134.786	40	108 - 137	30@10m
470 - 1000	40	960 - 1215	37@10m

Finally a warning is given in EN 55 011[6] clause 11 that ISM equipment is inherently capable of emitting EM radiation levels potentially hazardous to human beings and that it should be checked prior to testing with a suitable radiation monitor.

6.5 Summary

In this chapter the product-specific emission standard for information technology equipment, the generic emission standard for the 'residential, commercial and light industry' environment and their practical applications have been explicitly outlined together with the generic emission standard for the 'industrial' environment: EN 55 022[2], EN 50 081-1[3] and EN 50 081-2[4], respectively. The appropriate sections of the reference standards called up by the generic standards have also been reviewed. Key issues have been highlighted, principally:

- the need to have access to suitable test facilities and in most cases specialised instrumentation. This may involve companies in significant capital investment or the selection of a third party facility. In either case additional costs will be incurred.

- unlike many other types of electrical equipment testing, there is considerable uncertainty in the measurements made. Repeatability of testing is also a concern. Reasons for this are that:
  - — equipment configuration is open to interpretation;
  - — the calibration methods used for test facilities need further research;
  - — the generic standards apply to environment types, and therefore cover a broad range of products.
  - — for measurement methods the generic standards make reference to published standards and therefore have the same implications for manufacturers (*ie.* interpretational difficulties) as those apparent from the product specific standards.

- the generic standards allow a manufacturer flexibility to apply all or some of the tests specified, whilst this may be perfectly sensible it also allows for a variety of interpretation, such that for the same type of product it is possible that different tests may be applied by different manufacturers. This delegation of responsibility to a manufacturer may actually be regarded as consistent with the self-certification route to compliance with the EMC Directive, as the manufacturer is required to make his own declaration of conformity with the protection requirements of the Directive.

References

1. 89/336/EEC Council Directive 'on the approximation of laws of Member States relating to electromagnetic compatibility', Official Journal of the European Communities No.139, 25 May 1989, pp 19-26
2. EN 55 022: 1987 (BS 6527:1988) 'Limits and methods of measurement of radio interference characteristics of information technology equipment', British Standards Institution, 1988
3. BS EN 50 081 part 1 'Electromagnetic compatibility - generic emission standard generic standard class: residential, commercial and light industry', British Standards Institution, 1991
4 BS EN 50 081 part 2 'Electromagnetic compatibility - generic emission standard generic standard class: industrial', British Standards Institution, 1994
5. CISPR 22 'Limits and methods of measurement of radio interference characteristics of information technology equipment', IEC, 1985
6. BS EN 55 011: 1991 'Limits and methods of measurement of radio disturbance char-

acteristics of industrial, scientific and medical (ISM) radio-frequency equipment', British Standards Institution, 1991

7. Code of Federal Regulations No. 47, Volume 1 Part 15, FCC, 1993
8. BS 800: 1988 (EN 55 014: 1987) and AMD 6275 (No. 1 effective June 1990) 'Limits and methods of measurement of radio interference characteristics of household electrical appliances, portable tools and similar electrical apparatus', British Standards Institution, 1988
9. CISPR publication no.16 'CISPR specification for radio interference measuring apparatus and measurement methods', IEC, 1987
10. Bulletin OST55, Federal Communications Commission, Washington, August 1982
11. VDE 0877 part 2 'Measurement of radio interference Measurement of radio interference field strength', Feb.1985
12. EN 60 555 (BS 5406) part 2 'Disturbances in supply systems caused by household appliances and similar electrical equipment, part 2: Harmonics', British Standards Institution, 1988
13. EN 60 555 (BS 5406) part 3 'Disturbances in supply systems caused by household appliances and similar electrical equipment, part 3: Voltage Fluctuations', British Standards Institution, 1988
14. Amendment G to CISPR publication No. 22, IEC, October 1989
15. IEC 868 (1986) 'Flickermeter. Functional and design specifications', IEC, 1986 (Amendment No.1 1990)
16. 90/28296DC, draft revision of IEC 555-3 'Disturbances in supply systems caused by household appliances and similar electrical equipment. Part 3: Voltage Fluctuations', British Standards Institution, 1990
17. IEC 868-0 Technical Report 'Flickermeter Part 0: Evaluation of flicker severity', IEC, First edition 1991
18. 90/28293DC Draft IEC 555-5 'Disturbances in supply systems caused by household appliances and similar electrical equipment Part 5: Recommendations for voltage fluctuations in equipment exceeding 16 A or subject to special consent', British Standards Institution, 1990
19. IEC 50(161) 'International Electrotechnical Vocabulary Chapter 161: Electromagnetic compatibility', IEC, First edition 1990

7

Interpreting Immunity Standards

Chapter 7 is principally aimed at introducing the harmonised immunity standards to technical managers, design engineers, EMC test engineers and approvals engineers. The implications in terms of costs of testing and test facilities are also applicable to those controlling finances or marketing within manufacturing companies. An overview is given of the 'basic' standard, IEC 801 parts 2 to 4, referenced by the generic standard EN 50 082-1 and 2, also the proposed revision to IEC 801-3 and ENV 50140. The issues of test severity level and performance degradation are discussed. Hence the reader will acquire an understanding of immunity evaluation methods for radiated, conducted and ESD interference and an appreciation of the difficulties involved for these types of test, particularly interpretation of results.

7.1 Introduction

To enable self-certification for compliance with the EMC Directive [89/336/EEC[1]], a manufacturer must construct equipment to meet the requirements of the harmonised standards as discussed in Chapter 5. Not only must a manufacturer meet the emission requirements, he must also meet the immunity requirements.

Few commercial specifications exist for the measurement of conducted or radiated immunity. The US Mil-Std-461C[2] and DEF Std 59/41[3] specify immunity levels and measurement techniques for various categories of military equipment. Examples of standards which apply to commercial equipment are part of American National Standards Institute (ANSI) C63[4], SAMA PMC 33.1[5], IEC 801[6] and its BS equivalent, BS 6667: 1985.

The most important immunity standard is IEC 801[6], this now consists of 6 parts in various stages of agreement within the standards machinery. This standard was originally drafted by the IEC technical committee TC65 and covers the immunity of industrial-process measurement and control equipment. Part 1 is an introduction, Part 2 details evaluation of immunity to Electrostatic Discharge (ESD), Part 3 covers immunity to radiated electromagnetic energy and Part 4 conducted immunity to electrical fast transient/bursts. Currently Part 4 has not been adopted by BSI or CENELEC [HD 481[7]]. Part 5 covering disturbances in the power supply environment (surges) and Part 6 conducted interference *eg*. AM bulk current injection, have not been finalised by the IEC. Parts of this standard have also been adopted for testing other types of equipment, for example Part 3 was incorporated in CEGB Standard DN5[8] and now Parts 2, 3 and 4 are being called up as 'basic' or reference standards by the CENELEC generic immunity standard for the 'residential, commercial and light industry' environment.

In Chapter 5 the relevant immunity standards are listed, Tables 5.1 and 5.3, from which it is apparent that there is a shortfall of product specific immunity standards but this will be partly redressed by the introduction of generic standards. This problem is being tackled by the standards bodies but progress is slow. Examples of immunity standards will be considered to illustrate the problems faced by both the standards bodies and manufacturers who wish to self-certify their products. The examples chosen are IEC 801[6] Parts 2, 3, and 4 the 'basic' standards and the generic standard for the 'residential, commercial and light industry' environment, EN 50 082-1[9]. Reference will also be made to the draft documents prEN 55 101[10] Parts 2, 3 and 4 for information technology equipment (ITE) and the draft generic 'industrial' standard prEN 50 082-2[11].

7.2 IEC 801-2: 1984 First Edition and IEC 801-2: 1991 Second Edition Electrostatic Discharge (ESD)

7.2.1 Introduction to electrostatic discharge

Under certain atmospheric and environmental conditions, objects or people can become electrically charged. When two insulating materials with differing dielectric constants are rubbed together the materials become charged. One material gives up electrons to the other and the effect is described as *Electrostatic Charging*. By this process a person walking on a well insulated carpet in a dry atmosphere can become charged to several thousand volts. Upon approaching a conducting object the person is discharged *via* an arc. The discharge current and its associated high electromagnetic field can cause a malfunction or even destroy electrical systems such as computers or terminals.

This type of malfunction is common with information technology equipment, this is because heat dissipation from the system is given up to the surroundings which results in a large drop in the relative humidity, typically below 50%. There is a high risk that the operating personnel in this environment can become electrostatically charged, whether due to a synthetic carpet or clothing. If equipment is touched and the person discharged, the resulting disturbance may produce interference which shows up as a program error or loss of data (if the system is not adequately protected), although the person may only experience the sensation of a slight electric shock. Therefore it has become necessary for manufacturers of equipment to develop reproducible testing procedures to simulate the conditions found in practice for the discharge of a human body to equipment.

Figure 7.1 shows how voltage builds up between a human body and earth. Between steps a discharge takes place, after a few seconds a balance is achieved between charge and discharge rates; this is illustrated by Figure 7.1 and shows this voltage build-up graphically (plastic soles and polyamid carpet).

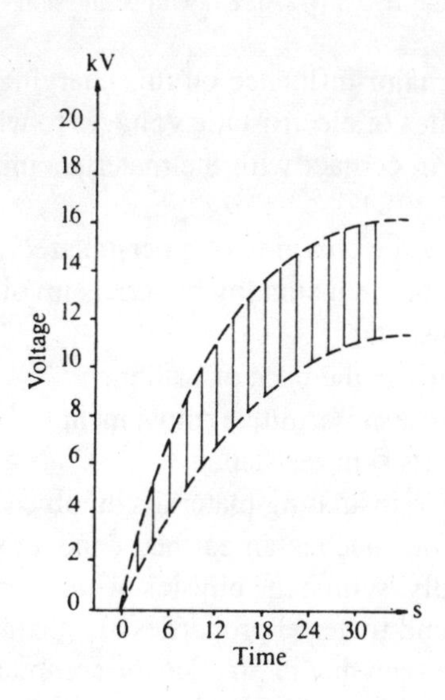

Figure 7.1 Graph of voltage build-up (plastic soles and polyamid carpet)

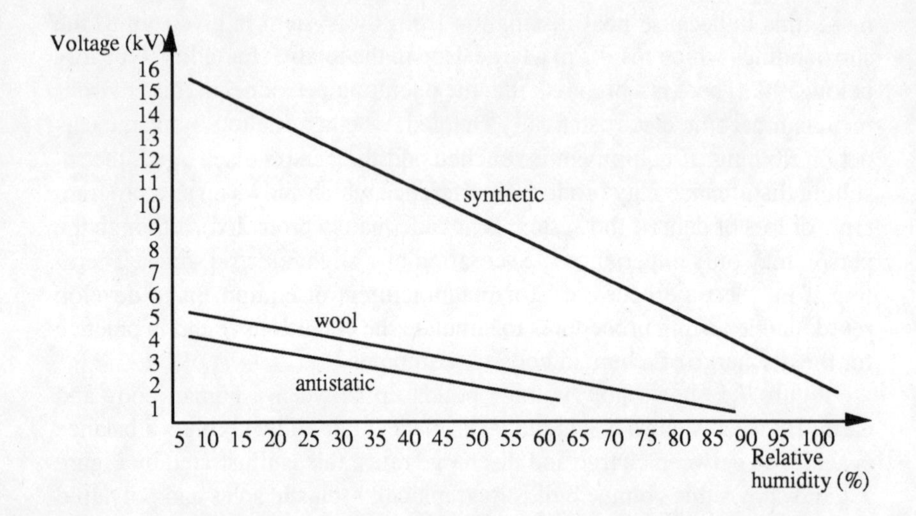

Figure 7.2 Maximum values of electrostatic voltages to which operators may become charged while in contact with the material indicated (IEC 801-2: 1984)

Reproduced from IEC 801-2: 1984, with permission

The environment has a major influence on this charging effect. Figure 7.2 shows the maximum values of electrostatic voltages to which operators' may become charged while in contact with the materials indicated. This is reproduced from IEC 801-2: 1984[6].

Other factors which affect the charging of a person are:

- the dielectric constants of the insulating materials involved (type of footwear, carpet, clothes *etc*)
- the rhythmic movement, *eg*. the pace of walking
- behaviour with time pauses/intermittent movement
- perspiration which affects skin resistance
- pressure between the two insulating materials involved.

When the charged person touches an earthed conductor then the discharge takes place very rapidly. Whilst the physics of electrostatic discharge (ESD) are well understood and material properties are characterised by the Triboelectric series, it can be seen that in practice the combination of factors in a given environment can be very complex and that the voltage prior to discharge and the discharge itself will be very variable. However, it is this discharge and the electrical equivalent circuit of a person that the standards for electrostatic discharge testing endeavour to replicate.

Two methods of ESD testing have been developed: Air Discharge and Contact-Discharge. The latter method is the result of recent work attempting to improve the repeatability of electrostatic discharge testing.

7.2.2 IEC 801 Part 2: 1984 (Air discharge)

ESD Generator — Figure 7.3 and Figure 7.4 show respectively the simplified diagram of the Electrostatic Discharge Generator and the typical output current (discharge current) reproduced from IEC 801-2: 1984[6]. The values of Rd and Cs mimic the resistances of the human body and its capacitance. The discharge current waveform represents the discharge current flowing from a person to an earthed conductor.

The Electrostatic Discharge Gun described in IEC 801-2: 1984[6] consists essentially of the circuitry shown in Figure 7.3 and a discharge electrode shaped to represent a finger [FIG 4, IEC 801-2: 1984[6]].

IEC 801-2: 1984 also specifies the characteristics and performance of the ESD generator [section 6.1, IEC 801-2: 1984[6]] and the

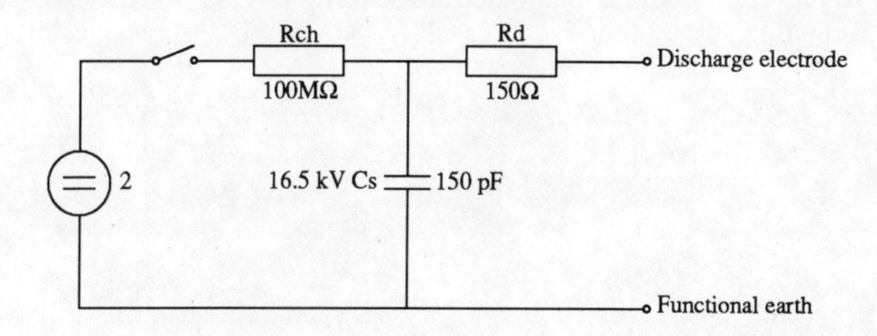

Figure 7.3 Simplified diagram of ESD generator
Reproduced from IEC 801-2 1984, with permission

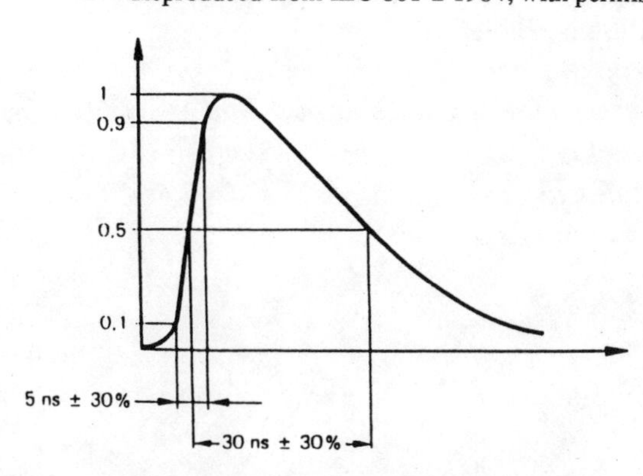

Figure 7.4 Typical waveform of discharge current
Reproduced from IEC 801-2: 1984, with permission

requirements for verification of the characteristics [section 6.2, IEC 801-2: 1984[6]]. The peak values of the discharge current should be:

9 A @ 2 kV,
18 A @ 4 kV,
37 A @ 8 kV and
70 A @ 15 kV, with a tolerance of +/- 30%.

The standard details a method for verifying the electrostatic discharge generator performance. The current waveform of Figure 7.4 is verified by discharging the gun to an 8 mm diameter electrode connected by a coaxial array of resistors representing a load of < 2 ohms, the waveform is measured by 100 MHz bandwidth oscilloscope *via* a 50 ohm matching resistor. Details of the verification test jig are given in FIG 2 of [IEC 801-2: 1984[6]]. There is now a number of laboratories accredited for performing ESD gun calibration.

Severity levels — The standard defines severity levels for ESD testing as shown in Table 7.1.

Table 7.1 Severity levels defined by IEC 801-2: 1984 for ESD testing

Level	Test voltage (+/- 10%)
1	2 kV
2	4 kV
3	8 kV
4	15kV

The appropriate severity level is selected by reference to Appendix A3 of the standard which suggests how the installation can be matched to the environmental conditions, Table 7.2.

Table 7.2 Severity levels and environmental characteristics

Class	Relative humidity as low as (%)	Antistatic	Synthetic	Max voltage (kV)
1	35	X		2
2	10	X		4
3	50		X	8
4	10		X	15

— for other materials class 2 is recommended *eg.*, wood, concrete, ceramic, vinyl, metal.

For acceptance tests the test programme and the interpretation of the test results are subject to agreement between the manufacturer and user, this must also necessarily include agreement on the environmental conditions under which the equipment will be used and therefore the severity level chosen for the equipment tests.

Test set-up — The equipment under test (EUT) is required to be placed on an earth reference plane and isolated from it by an insulating support about 0.1 m thick. This plane should be 0.25 mm thick copper or aluminium sheet (for other types of metal the thickness should be 0.65 mm) and a minimum of 1 m^2 depending on the EUT size, as the earth plane is required to project beyond the EUT by at least 0.1 m on all sides. The earth reference plane is connected to the earthing system.

The configuration of the EUT and its connections should be consistent with its normal functional requirements. A minimum distance of 1 m is required between the EUT and the walls of the laboratory and any other metallic structure. The EUT earth connection should be in accordance with the manufacturer's installation specification and no additional earths are allowed.

The ESD generator earth cable should be connected to the earth reference plane. This cable has a specified length of 2 m and in practice any excess cable should be bundled on the earth reference plane to avoid inductive loops.

The standard details the test set-up for table-top-mounted equipment, cubicles and systems in both the laboratory and field test conditions [FIGs 5, 6, 7, 8, 9, 10 of IEC 801-2: 1984[6]]. For example, Figures 7.5 and 7.6 illustrate the set-ups for table top equipment under laboratory and field conditions respectively. The field or *in situ* test set-ups may well be particularly useful to manufacturers of large equipment performing tests for inclusion in a Technical Construction File if they are using this route for demonstrating compliance with the EMC Directive.

Test procedure — The climatic conditions for performing ESD (Electrostatic Discharge) tests are specified as follows:

- ambient temperature: 15°C to 35°C
- relative humidity: 45% to 75%
- atmospheric pressure: 68 kPa to 106 kPa

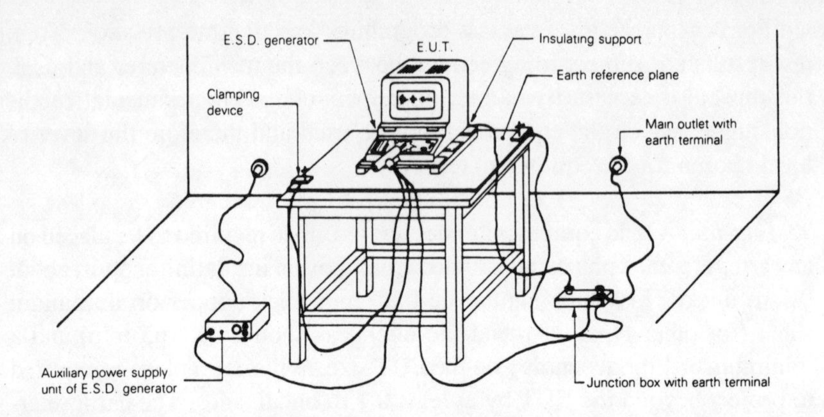

Figure 7.5 Test set-up for table-top-mounted equipment, laboratory tests
Reprodu~~d from IEC 801-2: 1984, with permission

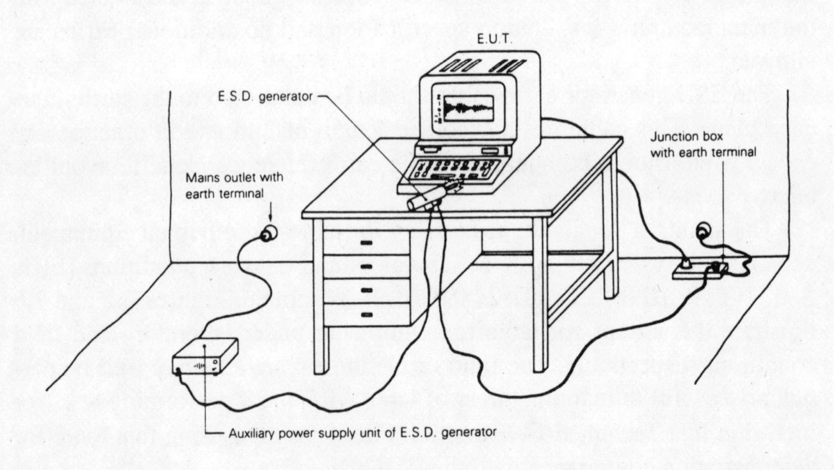

Figure 7.6 Test set-up for table-top-mounted equipment field tests
Reproduced from IEC 801-2: 1984, with permission

Direct application of the discharge to the EUT

The testing procedure is simple in principle. The ESD gun is charged to the appropriate voltage level and the discharge electrode is then pointed at the various parts of the equipment that it is anticipated an operator will touch in normal usage. The ESD generator is held perpendicular to the surface to which the discharge is to be applied and the electrode is moved towards the EUT until discharge occurs. The amplitude of the discharge should be gradually increased from minimum to the selected test severity level; this should not be exceeded in order to prevent damage to the EUT. The test should be performed with single discharges on the pre-selected points and at least 10 discharges should be applied to each point with an interval of at least 1 s between them.

In order to establish the points to which the discharges should be applied pre-testing may be required and a discharge repetition rate of 20 discharges per second is permitted by the standard. Discharges to points on the EUT which are accessible for maintenance only are not required unless agreed by the manufacturer and user.

Indirect discharges in the vicinity of the EUT
Discharges to objects which may be installed near the EUT are simulated by applying discharges to the earth reference plane. This method is also used for equipment in non-metallic housings. At least 10 single discharges should be applied on each accessible side of the EUT, with the generator at least 0.1 m from the EUT.

Evaluation of results — IEC 801-2: 1984[6] does not specify EUT performance degradation but simply suggests effects which should be recorded and states that, 'in the case of acceptance tests the test programme and the interpretation of the test results are subject to agreement between manufacturer and user'. The test documentation should include the test conditions and the results. These results should include:

- the effect of ESD on the EUT output as a consistent measurable effect
- the effect of ESD as a random effect occuring only during the application of the discharge
- the effect of ESD as a random effect lasting after the discharge but which can be 'reset'
- any damage occuring to the EUT as a consequence of applying the discharge

Whilst this approach is acceptable where a manufacturer is required to perform type tests for a user, or needs to demonstrate that the user's specifications have been met, it is not so helpful to manufacturers attempting to comply with the EMC Directive as it is left to the manufacturer to decide what is an acceptable level of performance degradation. However the generic standards do specify performance criteria which can be applied, see section 7.5.

7.2.3 IEC 801-2: 1991 part 2 (Contact discharge)

The Air Discharge method of ESD testing described in IEC 801-2: 1984[6] has been demonstrated to have a number of shortcomings. The reproducibility has been shown to be dependent on:

- the speed of approach of the discharge tip, humidity, and the construction of the test equipment; leading to variations in the rise time and the magnitude of discharge current.

this resulted in a situation where:

- the tests performed did not always simulate field failures.
- equipment could fail at low voltage test levels, yet pass at higher levels.
- the test procedure for simulating discharges to objects in the vicinity of the EUT was not well defined, yet this test was very important for EUT's having plastic cases.
- the categorisation of the effects of the test on the EUT were not well defined.

[B Jones, 1990[12]].

Results from these investigations have led to an alternative method of carrying out ESD testing, the 'Contact discharge' method. This method is now described in IEC 801-2: 1991 Second edition[13] and is also adopted by the draft prEN 55 101-2[10]. It should be noted that these standards still include the air discharge method.

Electrostatic discharge generator — Figure 7.7 shows the simplified diagram of the ESD Generator and when compared with Figure 7.3 it can be seen that Rd, the resistance of the human body, has been increased to 330 R and a 'discharge switch' introduced. The reasons for these changes are shown by Figure 7.8 — the discharge current waveform. When compared with Figure 7.4 the overall envelope is similar in shape, ignoring the high initial spike. It is now recognised that this initial spike has always been present but was concealed by the bandwidth limitations of the measuring equipment available a few years ago. For the air discharge method the immunity level for the equipment was defined as a function of electrostatic voltage, however it has been shown that the energy transfer is a function of the discharge current not the voltage prior to discharge. Further it has been found that the discharge current is not proportional to the pre-discharge voltage at higher pre-discharge levels [Jones, 1990[12] and IEC 801-2: 1991[13]].

To overcome the variation in rise time, due to speed of approach of the discharge tip, the 'discharge switch' has been introduced into the ESD generator circuit, Figure 7.7. In practice this switch is a vacuum or mercury wetted relay contact, which produces repeatable fast rising discharge currents. Prior to discharge Cs is charged to the defined discharge voltage. When the relay is closed there is an initial fast rising pulse of current, followed by the discharge of Cs through Rd into the EUT. The testing procedure differs from the air discharge method by having the discharge tip actually in contact with the EUT prior to closing the relay contacts.

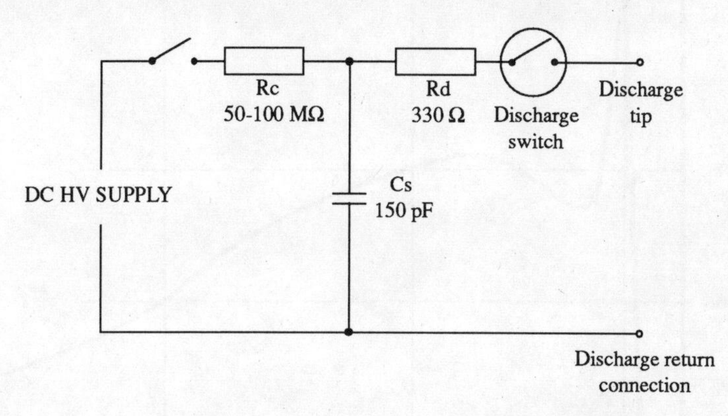

Figure 7.7 ***Simplified design of revised ESD generator***

A specification for the characteristics and performance of the ESD generator are defined in section 6.1 of IEC 801-2: 1991[13] and verification of performance requirements in section 6.2 of IEC 801-2: 1991[13]. The verification details include the discharge current characteristics and constructional details for the 'current sensing transducer' [Annex B of IEC 801-2: 1991[13]]. The minimum recommended bandwidth for the measurement oscilloscope is 1GHz.

Severity levels — Severity levels are specified for both the contact and air discharge methods of ESD testing. The air discharge limits are as defined in IEC 801-2: 1984[6], (*see* Table 7.1), whilst the contact discharge severity levels are as shown in Table 7.3.

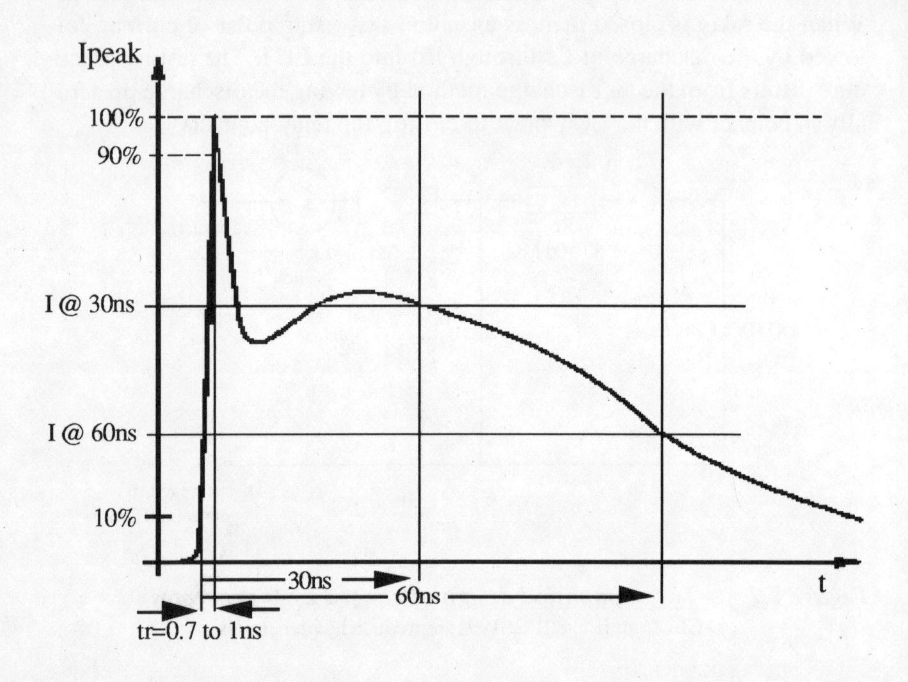

Figure 7.8 Typical waveform of the output current of the ESD generator [IEC 801-2: 1991]

Table 7.3 ***Severity levels defined by IEC 801-2: 1991 for 'contact discharge'***

Level	Test voltage (+/- 10%)
1	2 kV
2	4 kV
3	6 kV
4	8 kV
x	Special

Level 'x' is defined as being subject to negotiation and should be specified in the dedicated equipment specification. Annex A4 of IEC 801-2: 1991[13] contains examples of the application of the severity levels related to environmental or installation classes. These examples are identical to those included in IEC 801-2: 1984[6] and shown in Table 7.2.

Contact discharge is the preferred method and air discharge should only be used when contact discharge cannot be applied.

Test set-up — The test set-ups described in the standard are similar to the earlier method [IEC 801-2: 1984[6]], except for the distinction that contact discharge is made to conductive surfaces and coupling planes, whereas the air discharge is used for insulating surfaces.

Two types of test are included:

- type tests performed in a laboratory, this is the preferred method and only method which may be used for claiming conformance with the standard
- *in situ* or post-installation tests, performed on equipment in its final installed conditions

The test set-up and equipment configuration is then as specified in the earlier version of the standard already described in section 7.2.2, excepting for the provision of coupling planes for indirect discharges.

The horizontal coupling plane (HCP) and vertical coupling plane (VCP) should be made from the same material and be of the same thickness as the earth or ground reference plane (GRP). The coupling planes are connected to the GRP via resistive cables. These have 470 kohm resistors at each end of the connecting cable.

Figure 7.9 [Figure 5, IEC 801-2: 1991[13]] shows the set up for a table-top EUT. A wooden table is placed on the ground reference plane (GRP), the horizontal coupling plane (HCP) is placed on the table and the EUT and associated cables are placed on the HCP but isolated from it by 0.5 mm thick insulation. The HCP is defined as being 1.6 m x 0.8 m and the EUT is required to be a minimum of 0.1 m from all sides of the HCP. If required two HCPs 0.3 m apart with their short sides adjacent may be used, in this instance the HCPs are not bonded together but are independantly connected to the GRP *via* resistive cables. The VCP is 0.5 m x 0.5 m and is positioned 0.1 m away from and parallel to the EUT. The VCP is moved during testing so that each of the four faces of the EUT are illuminated in turn.

Test set-ups are illustrated in the standard for table-top equipment [Figure 5, IEC 801-2: 1991[13]], floor-standing equipment [Figure 6, IEC 801-2: 1991[13]] and for post-installation tests. This latter *in situ* method may be useful during the preparation of technical construction files for some types of equipment or may be called up by future standards, this was noted also for the earlier edition of the standard. A portable GRP 2 m x 0.3 m is placed on the floor 0.1 m away from the installation and connected to the protective earthing system where the installation allows, or to an earthing terminal on the EUT.

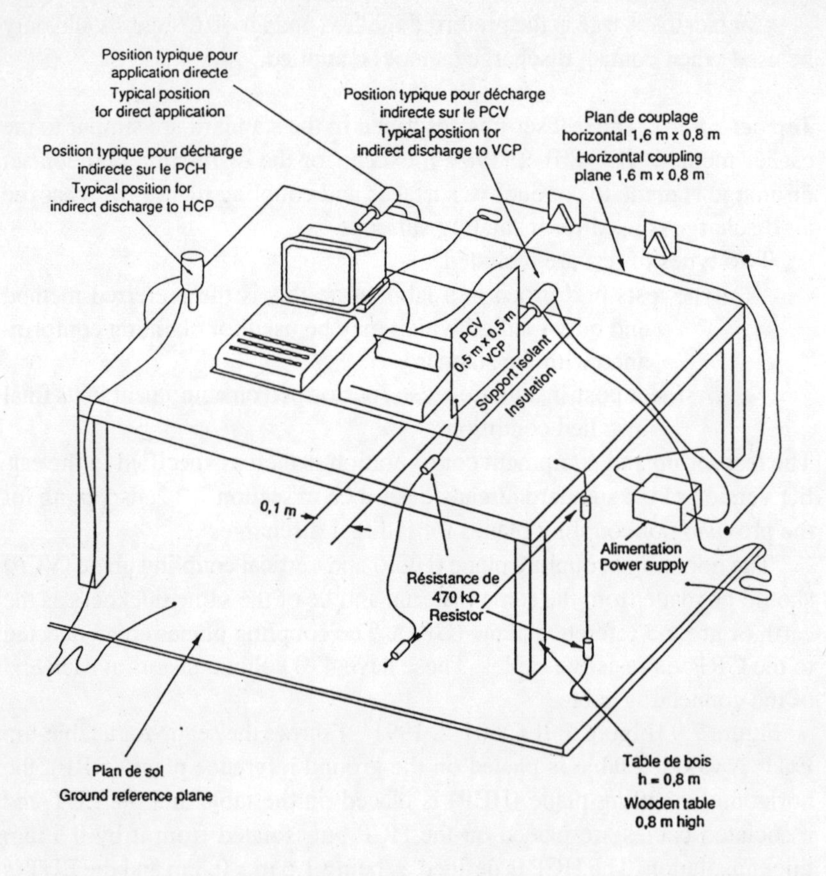

Figure 7.9 Example of laboratory test set-up for table-top equipment

Test procedure — The climatic conditions during testing are as specified for the first edition of the standard (*see* 7.2.2).

During testing it is essential that the EUT is exercised in all its normal modes of operation and any test software or programmes should reflect this. For conformance tests the EUT is to be operated in its most sensitive mode which should be determined by preliminary testing. Any monitoring equipment that is used should be adequately decoupled to prevent erroneous failures.

It is recommended that a test plan be drawn up which should establish the following parameters:

- representative operating conditions
- type of EUT, *ie*. table-top or floor standing
- indirect application to HCP and VCP, and positions of VCP
- points at which discharges to be applied
- type of discharge at each point, contact or air discharge
- the number of discharges at each point.

Some aspects of the test plan may only be determined by performing investigatory pre-tests. Examples of locations affecting the choice of points to which discharges should be applied are:

- metallic sections of cabinets electrically isolated from ground
- the 'man-machine interface' *ie*. the control/keyboard area -switches, knobs, buttons or other operator accessible areas
- indicators, LEDs, slots, grilles, connector hoods, *etc*

Direct discharges should only be made to those points and surfaces accessible to personnel during normal usage, including maintenance, but not to those areas accessible for maintenance only. The ESD generator should be held perpendicular to the discharge point/surface and the test voltage should be gradually increased from minimum to the selected severity level to determine the threshold of failure. Ten single discharges should be applied in the most sensitive polarity on the pre-selected point, allowing at least 1s between discharges. The discharge points may be selected by exploration using a repetition rate of 20 discharges/s.

For contact discharges the tip should be in contact prior to the discharge switch being closed. In the case of painted metal surfaces, the surface should be punctured by the pointed tip. Where such a surface is intended to be insulating it should be subjected to air discharges.

For an air discharge the discharge switch is closed and the rounded tip used. The tip should approach the EUT as fast as possible consistent with not actually causing mechanical damage when the tip touches the EUT.

Discharges to objects placed or installed near the EUT are simulated by making contact discharges to the coupling planes. At least ten discharges should be made to the HCP at points 0.1 m from each side of the EUT in the most sensitive polarity. Ten single discharges should be made to the centre of one vertical edge of the VCP, with the VCP in the positions defined in the test plan.

Evaluation of test results — As for the earlier edition of the standard the variety and diversity of equipment to which it could apply makes general evaluation criteria difficult to define. However a classification of the test results is suggested based on the following criteria:

1) normal performance within the specification limits
2) temporary degradation or loss of function or performance which is self-recoverable
3) temporary degradation or loss of function or performance which re quires operator intervention or system reset
4) degradation or loss of function or performance which is not recover able, due to damage of equipment (components) or software, or loss of data.

The standard also states 'in the case of acceptance tests, the test programme and the interpretation of the test results are subject to agreement between manufacturer and user'.

The requirements of IEC 801-2: 1991[13] have also been adopted by prEN 55 101-2[10] which is the draft electrostatic discharge standard for ITE (likely to become EN 55 024-2), it has also been adopted by CENELEC as EN 60801-2: 1993, the transposed British Standard being BS EN 60801-2: 1993. Finally it will be adopted by the IEC as a basic standard in the IEC 1000 series, IEC 1000-4-2.

7.3 IEC 801-3: 1984 (BS 6667: 1985), 90/29283 DC - Draft Revision of IEC 801-3, 1990 and ENV 50140: 1994.

7.3.1 IEC 801-3: 1984 (BS 6667: Part 3: 1985) General

IEC 801: Part 3: 1984[6] describes the measurement methods and defines the levels of immunity to radiated electromagnetic fields for industrial-process measurement and control equipment. In particular it seeks to 'establish a common reference for evaluating the performance of industrial-process measurement and control equipment when subjected to electromagnetic fields such as those generated by portable radio tranceivers (walkie-talkies) or any other device that will generate continuous wave radiated electromagnetic energy'.

The current issue of the standard will now be discussed as it illustrates how a standard can prove inadequate and also how it can be misinterpreted. This is particularly important as this standard is called up by the generic immunity standard for the 'residential, commercial and light industry' environment, EN 50 082-1[9] (section 7.4). The proposed revision to IEC 801-3 [90/29283, 1990[14]] and subsequent proposals (ENV 50140[21]) will also be reviewed, illustrating both its implications for industry and providers of EMC testing and the evolution of standards.

The object of radiated immunity testing is to illuminate the EUT with electromagnetic energy at a prescribed field strength.

Test methods — The test methods described in IEC 801-3[5] allow the use of shielded enclosures (screened rooms), anechoic chambers and the 'stripline circuit'. In Annex A of IEC 801-3[5] the TEM cell, mode-stirred chamber and open antenna ranges are also referred to. The latter method however, can be disregarded since the transmitted energy levels required for the immunity testing are likely to be in excess of the permitted legal limits. Further, the local environment should be shielded from the relatively high EM fields which are generated during testing in order to minimise the potential for localised interference. The test and monitoring equipment will also require shielding so this is usually situated outside the screened enclosure. In effect the screened enclosure becomes a barrier protecting the local enviroment from the potential elctromagnetic disturbances which will be generated during testing.

When using the screened room or chamber, the equipment under test (EUT) is illuminated with electromagnetic energy of the appropriate field strength by using antennas positioned 1 m away from the EUT. The field strength is measured by placing field probes adjacent to or on the EUT. Screened rooms are known to have shortcomings due to standing waves and reflections and there is therefore the potential for erroneous results. However IEC 801-3: 1984[6] provides a rationale for justifying the use of shielded enclosures and concludes that they offer the 'most efficient means of performing radiated measurements that is in *widespread use*'. It is suggested that the wide excursion of field strengths can be overcome by the use of one or more field probes placed near the EUT. As the peaks and nulls are encountered the field strength can be readjusted to the required level. In practice standing waves and reflections are also usually controlled by the introduction of some radio absorbent material (RAM) giving a degree of partial anechoic shielding.

Included in note 2 to section 6 of IEC 801-3: 1984[6], is a reference to the substitution method of field strength measurement. For this method the field strength is measured using an appropriate field probe within the volume to be occupied by the EUT and then the EUT substituted for the probe. The standard contends that this method is 'prohibitively expensive' and its merits questionable since the EUT will distort the original field measured. The results of current work in this area are contrary to this contention and indeed this is reflected in the proposed revision to IEC 801-3 [90/29283[14]] and the draft radiated immunity standard for ITE [prEN 55 101-3[10]]. It is noted in IEC 801-3: 1984[6] that anechoic chambers will substantially reduce errors above 100 MHz and that this is the 'preferred type of enclosure'.

A useful method for small EUTs is to use a stripline circuit (Figure 7.10) within a screened enclosure. When the stripline circuit is used the EUT is placed in the centre of the 'cube' part of the stripline on a support of foam. The aim of the stripline is to provide a uniform electric field within the cube position. The parallel stripline specified has dimensions of 80 cm x 80 cm x 80 cm and it is suggested that this is usable up to 500 MHz. The EUT size is limited to one-third of the dimensions of the stripline for larger equipment otherwise the field is likely to be perturbed. Full constructional details of the stripline are given in FIGs 2 to 5 of IEC 801-3: 1984[6].

The recommended test equipment is:

1) a shielded room or anechoic chamber having an adequate size to maintain the distances specified in Figure 7.9
2) a stripline circuit (27 MHz to 500 MHz), Figure 7.10
3) a signal generator capable of covering the frequency range and having automatic sweep capability of 0.005 octave/s (1.5×10^{-3} decades/s) or slower
4) a power amplifier to amplify the signal generator output (when necessary) and drive the transmitting antenna (or stripline)
5) Transmitting antennas: biconical, 27 MHz to 200 MHz; conical logarithmic spiral, 200 MHz to 500 MHz; or 'any other antenna system capable of satisfying the frequency requirements'
6) isotropic field strength probe(s)
7) any associated monitoring equipment.

Severity levels — Radiated immunity levels and measurement techniques are specified by this standard. For severity levels 1 to 3 and x the test field strength is specified in Table 7.4

Table 7.4 Test severity levels IEC 801-3: 1984

Level	*Test field strength V/m*
1	1
2	3
3	10
x	Special

Frequency Band 27 to 500 MHz

The problem faced by the manufacturer is of choosing an appropriate level of severity to test his equipment against. Here it is necessary to consider the threat level likely to exist within the operating environment for the equipment. The manufacturer also has to consider the level of performance degradation which can be tolerated. Here there may be differences in what is considered acceptable between the manufacturer and the end user. For example, whether a temporary flicker on a display is acceptable? It should be noted that severity level x is determined by agreement between manufacturer and user.

Annex A9 of IEC 801-3: 1984[6] provides general guidelines to the choice of severity level. The classifications are as follows:

Class 1: Low level EM environment, such as levels typical of local radio/television stations located at more than 1 km and levels of power typical of low power transceivers.

Class 2: Moderate EM environments, such as portable transceivers that can be relatively close to the equipment but not closer than 1 m.

Class 3: Severe EM environments, such as levels typical of high power transceivers in close proximity to the control equipment.

Class 4: Open class for situations involving very severe EM radiation environments. The level is subject to negotiation between the user and manufacturer or as defined by the manufacturer.

For manufacturers using the generic standards the severity level is defined by the appropriate standard, *eg*. for the residential, commercial and light industry environment the field strength specified is 3 V/m.

Test set-up and procedure

Shielded room

The EUT is placed on a wooden table in the centre of the enclosure. If the EUT is a rack-type of equipment it is placed on the floor but insulated from it. If the EUT or its cabling requires support then this should be constructed from non-metallic material. The earthing of the EUT should be consistent with the manufacturer's installation recommendations.

The radiating antenna should be placed at least 1 m from the front of the EUT.

All testing should be performed in conditions representing as closely as possible the normal installation conditions; this means both the configuration and wiring layout of the EUT. If the interconnecting wiring is unspecified then an unshielded twisted pair should be used which will be exposed to the incident EM radiation for a 1 m length from the connection to the EUT. After this the wiring may be filtered and screened if it is connected to test equipment outside the screened room.

The required field strength is measured by placing the isotropic field probe(s) on top of or directly alongside the EUT and monitoring the field strength *via* a remote field strength indicator outside the enclosure. This enables the field strength to be adjusted by controlling the RF amplifier.

The test should be performed with the antenna directed at the most sensitive side of the EUT. For the biconical antenna, measurements will be required for both vertical and horizontal polarisations, whilst for the log spiral its circular polarisation means that no change in the antenna position is required.

The frequency range is swept from 27 MHz to 500 MHz. It will be necessary to pause to adjust the field strength and change antennas. The rate of sweep is in the order 1.5×10^{-3} decades/s. For frequencies at which the EUT appears to be sensitive then this should be investigated by setting the signal generator to the particular frequencies and adjusting the field strength to determine the threshold for the degradation of performance.

Stripline circuit (Figure 7.10)
An EUT having dimensions of 25 cm x 25 cm x 25 cm or less may be tested in the stripline specified. The EUT is placed in the centre of the cube part of the stripline supported on foam plastic and is tested in three different orientations. As for the shielded room the equipment configuration and wiring should represent as closely as possible the conditions found in practice.

The filters on the top of the stripline are intended to protect the signal and power lines to the EUT and reduce any conducted interference liable to affect the external test instrumentation.

To prevent reflections the stripline is placed 2 m away from walls or any metallic enclosure. The power and signal lines to the EUT are connected *via* the filters on the top of the stripline. Outside the stripline these leads are routed vertically for 0.5 m and then horizontally to any associated equipment which is positioned 2 m away on the axis of the stripline.

The required field strength is measured by reading the voltage between the two plates. This is converted into field strength *via* a calibration factor determined by measuring the field with an isotropic field probe located in the EUT position.

The test is performed in the same way as for the test within a shielded enclosure.

Evaluation of test results — IEC 801-3: 1984[6] does not specify an acceptable performance degradation for the EUT. It attempts to classify the effects as follows:

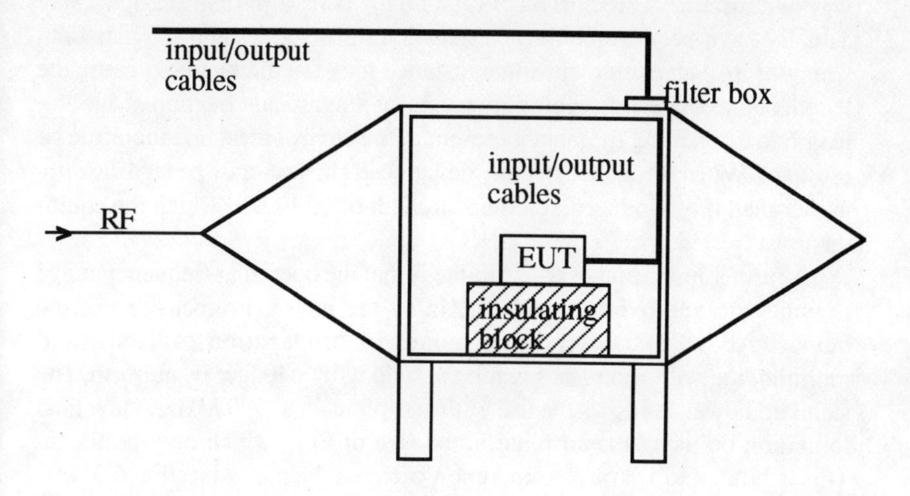

Figure 7.10 ***Test set-up in the stripline circuit (IEC 801-3)***

a) The effect of the EM field on the EUT output,
   i) as a consistent measurable effect,
   ii) as a random effect, not repeatable, either as a transient effect occuring during the application of the EM field or as a permanent or semi-permanent effect after the application of the EM field.

b) Any damage to the EUT resulting from the application of the EM field.

As with IEC 801-2: 1984[6], acceptance tests, test programme and interpretation of results are to be agreed between manufacturer and user and the test documentation should include the test conditions and the results. For manufacturers using IEC 801-3: 1984[6] as part of conforming with the generic standards, then the criteria for performance degradation is better defined, *see* 7.5.

Discussion — The stripline method can at best be described as giving a repeatable test result. However, it does not guarantee that the equipment will perform correctly when installed in its working environment.

In practice the stripline described in the standard is found (depending on the input impedance of the EUT) to be resonant in the frequency range 50-60 MHz and hence the induced currents in the EUT connecting cables are significantly higher in this frequency band. Therefore, an EUT tested in this way and meeting its requirements at a nominal 10 V/m field strength, may actually have been tested at the equivalent of a significantly higher field strength. Also an equipment when installed may fail in practice because the length of the connecting cable may represent a significant fraction of a wavelength to an exciting frequency present in the environment and therefore be resonant. When resonance occurs the induced currents may be significantly higher than those induced at a field strength of 10 V/m at which the equipment was tested.

A further limitation of the stripline is that the operating frequency range is limited to approximately 200 MHz by the physical dimensions of the device. Above 200 MHz other modes of propagation can exist and multimoding will occur. As a result the field will no longer be uniform. The standard however allows the use of the stripline up to 500 MHz. A practical limitation on using the stripline is the size of EUT which may be tested. This is limited to 1/3 of the separation between the plates [section 6.3, IEC 801-3[6]]. The maximum size of equipment which can be tested is therefore limited to 25 cm x 25 cm x 25 cm [section 8.2, IEC 801-3[6]].

It is apparent that the standard itself is open to misinterpretation. This is particularly so in respect of the choice of severity levels to be used for testing and also the limitations on the test methods described. This can be illustrated by the practical experience of using multiplexing equipment in a petrochemical plant. The equipment was found to meet the requirements of IEC 801-3[6] severity level 3, but was found to maloperate in a particular installation. It may therefore be essential for some manufacturers to carry out a threat evaluation in a particular environment before selecting a severity level for the tests. The standard must be used with caution, since the EMC Directive requires compliance with the 'essential protection requirements' [Article 4, 89/336/EEC[1]], not with a particular standard.

Immunity standards are coming under scrutiny as the IEC and CENELEC strive to produce standards which can be effectively used and, identified deficiencies are being addressed. In the case of IEC 801-3 a revision has been published in draft form. The proposed revisions to IEC 801-3 are described in the BS draft document 90/29283[14], entitled: 'Revision of IEC Publication 801-3: Electromagnetic Compatibility for Industrial-Process Measurement and Control equipment Part 3: Immunity to radiated radio frequency Electromagnetic Fields.' Most of the proposed changes were also included in the proposed radiated immunity standard for ITE, prEN 55 101-3[10] (which will be EN 55 024 when adopted). Because these changes were still being debated in 1994, CENELEC have issued ENV 50140[21] which is a'draft for development' but which is an official document which can be referred to.

7.3.2 90/29283 DC Draft - revision of IEC publication 801-3: electromagnetic compatibility for industrial process measurement and control equipment Part 3: immunity to radiated radio frequency electromagnetic fields and ENV 50140: 1994

This draft revision of IEC 801-3 was circulated for comment in September 1990. Compared with the 1984 issue of IEC 801-3[6] some major changes are proposed in respect of test methods. In particular, the stripline circuit is only listed in the alternative test methods and its frequency limitations are recognised. The antenna to EUT spacing within a screened enclosure is increased to 3 m and the generated field is to be amplitude modulated by a 1 kHz tone. More fundamentally the standard has been identified as a 'Basic' EMC publication by the IEC's Advisory Committee on EMC (ACEC) in accordance with IEC Guide 107[15] and although the title on the discussion document refers to 'industrial-process measurement and control equipment' the draft IEC text simply refers to 'electrical and electronic equipment', this will be adopted as IEC 1000-4-3. Various revisions to the revision document have also been proposed resulting in the CENELEC ENV 50140[21].

General — Radiated immunity testing is required to evaluate the immunity of equipment to radiated EM fields present in the operating environment. IEC 801-3: 1984[6] specifically refers to the threat from transceivers (walkie-talkies) and the revised standard, in line with its new remit as a basic standard, broadens the scope of the potential threat to include spurious EM energy which may be radiated from, for example, arc welders, thyristor drives, fluorescent lights and switches operating inductive loads.

The EM environment is determined by the strength of the field measured in volts/m. It is noted that this is not easily measured without sophisticated instrumentation neither can it be easily calculated because of the effects of surrounding structures or proximity of other equipment which will distort or reflect the EM waves.

Essentially, the EUT is illuminated with radiated electromagnetic energy. This should simulate the conditions found in practice, and should allow the performance degradation of the EUT to be monitored. The electromagnetic fields generated can be either modulated or unmodulated. The test methods defined are intended to establish adequate repeatability of results at different test facilities for qualitative analysis of effects.

The following will be considered in detail:

the test facility
the severity levels
the field strength uniformity

test set-up and procedure
evaluation of test results

Test facility — Radiated immunity testing requires the generation of specified electromagnetic fields. However, this can be problematic because generated fields of this kind may interfere with local radio services or other electrical equipment being used nearby and may infringe national or international laws. This practical difficulty is recognised in the draft revision which suggests that testing should be performed in a shielded enclosure. Also because sensitive test instrumentation is used this should be shielded from the generated EM fields.

The preferred facility is an anechoic chamber or absorber-lined shielded enclosure, together with auxiliary screened rooms to accommodate the field generating and monitoring equipment and the equipment required to exercise the EUT. The size of the absorber lined shielded enclosure should be suitable for the size of the EUT to be tested, in order to achieve adequate control over the tolerance on the required field strength. The standard states that the antenna to EUT distance should be greater than 1 m, however 3 m is preferred and the test report should indicate if the test distance is less than 3 m. In cases of dispute, measurements at 3 m will take precedence. No further guidance is given on the dimensions of the enclosure. prEN 55 101-3[10] the draft radiated immunity standard for ITE does indicate certain minimum dimensions, these are shown in Figure 7.11. However, the overriding criterion is the uniformity of the field strength that is illuminating the EUT. Larger or smaller screened enclosures may be used if it can be demonstrated that the radiated field is not altered in such a way as to affect the test results. Anechoic chambers are less effective at lower frequencies because the absorber becomes less efficient, typically at frequencies below 80-100 MHz. Therefore greater care must be used to establish the uniform field and extra RF absorber may be needed, positioned by experiment, to have the greatest effect. Therefore ENV 50140[21] suggests that the lower frequency limit be raised to 80 MHz and that for lower frequencies 'bulk current injection' techniques should be used as described in the draft IEC 801-6 and ENV 50141.

Alternative methods of generating the required EM fields are permitted, including TEM cells and stripline circuits, unlined screened rooms, partially lined screened rooms and open antenna ranges. As previously discussed the latter cannot really be considered and the other facilities all have limitations. The TEM cell and stripline are both limited by the usable frequency range, typically up to around 200 MHz and the size of EUT which can be accommodated.

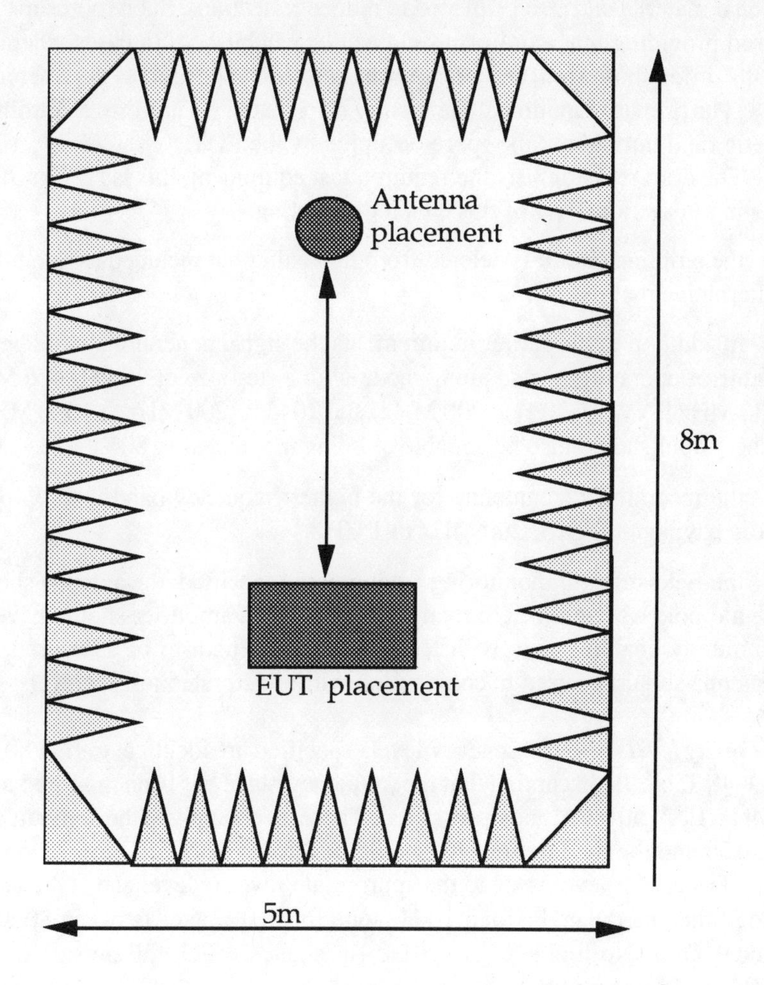

Notes:
- Absorber material pyramid height = 1 m approx.
- Ceiling is also covered with absorber material of the same size as the walls.
- Absorber lined chamber dimensions are based on maintaining 1 m spacing from any point of antenna or EUT to closest point of absorber.

Fig 7.11 Absorber lined screened enclosure (dimensions based on prEN 55 101-3)

The TEM cell being enclosed may be used in a normal laboratory environment but the stripline should be used within a screened enclosure. Unlined screened room resonances can be damped by the addition of absorbing materials positioned at strategic points (*see* Chapter 8 and Appendix F), additional material can also be placed to reduce reflections. Such rooms may be used providing that a uniform field can be established. In cases where results differ, those from an anechoic chamber are to be used as the reference.

The primary function of the facility is to generate a uniform electromagnetic field throughout the space occupied by the EUT.

The draft revision lists the required test equipment, this is similar to the listing in section 7.3.1 of this chapter excepting:

- the stripline circuit is deleted from the listing but included as one of the alternative methods

- in addition to the earlier requirements the signal generator should be capable of operating in a stepping mode with a step size of 10 kHz (26 MHz [80 MHz ENV 50140[21]] to 200 MHz) and 20 kHz (200 MHz to 1000 MHz). The output should also be capable of being modulated to 80%.

- the recommended antenna for the higher frequency band is a log-periodic having a range of 200 MHz to 1 GHz.

- the field strength monitoring equipment is specified: the antenna should be a dipole having a 10 cm total length, the head amplifier should have an immunity enabling it to withstand the field strength to be measured, the antenna should be used in conjunction with a calibrated test receiver.

Severity levels — The severity levels specified are identical to IEC 801-3: 1984[6], Table 7.4 excepting that the frequency range has been modified to 80 MHz [ENV 50140[21]] and up to 1 GHz. These levels apply to the unmodulated carrier and the field uniformity should be 0 to +6 dB.

The carrier level is set to the appropriate severity level and is then 80% amplitude modulated with a 1 kHz tone for frequencies between 80 MHz and 1 GHz. No limits are specified for signals which fall outside the 80 MHz to 1 GHz range.

Advice on the selection of severity level is given in Annex A5 to the draft revision [90/29283[14]]. The severity classes are similar to those already listed for IEC 801-3: 1984[6], with the following modifications:

Class 1: No change.

Class 2: Moderate EM environment. Low power portable transceivers in use, typically less than 1 W rating, having restrictions on use when in close proximity to the equipment. A typical commercial environment.

Class 3: Severe EM environment. Portable transceivers (2 W rating or more) are in use relatively close to the equipment but not less than 1 m.

High power broadcast transmitters or ISM equipment in close proximity. A typical industrial environment.

Class 4: No change.

Uniformity of field strength — The uniformity of the field strength is measured by placing the field sensing probe at the EUT position and adjusting the carrier signal level of the transmitting system until the required field strength is obtained. This procedure is repeated for other sensor positions within the 1.5 m x 1.5 m 'area' intended for the position of the front face of the EUT to ensure uniformity of the generated field. A field is considered to be uniform if its magnitude does not vary by greater than -0 to +6 dB of the nominal value, over 80% of the surface [NB the tolerance on field uniformity is not specified by ENV 50140!].

This tolerance is expressed in this way in an attempt to ensure that within the test volume the field strength does not fall below the nominal. It is considered by the standard that this tolerance is the minimum that can be achieved in a practical test.

The substitution method of field measurement is now considered to give more repeatable measurements than the method described in IEC 801-3: 1984[6] where the field probes are required to be placed on or adjacent to the equipment. Results of research published in 1992, Dawson, Mann and Marvin[16], confirm this. Of further concern is the fact that, although the standard allows any facility within which the required field uniformity can be achieved, in practice this may actually mean anechoic or semi-anechoic chambers and further work is required to establish the minimum requirements for a facility.

Test set-up and procedure — The EUT should be tested under conditions as close as possible representing those found when it is installed or in use. Wiring and cabling should also be consistent with the manufacturer's recommended procedures.

Cabling between enclosures of the EUT should be treated as follows:

• the manufacturer's specified cable types and connectors should be used

• if the specified length is less than or equal to 3 m, the specified length shall be used

• if the length is greater than 3 m, then it should be reduced to 3 m by non-inductive bundling

• the cables should be arranged within the uniform volume of the field to minimise immunity. This may require investigation, but generally this can be achieved by aligning the cables, as far as possible, in the axis of the incident field. (Although not stated presumably the requirement to meet normal installation conditions takes precedence over this arrangement of cabling?)

As in the earlier version of the standard where wiring to the EUT is unspecified, a 1 m length of unshielded cable should be connected to the EUT. However in the proposed revision parallel conductors are specified, whereas in the original version twisted wires were specified. These should be connected *via* filters and shielded cables to the exercising equipment outside the shielded enclosure. The exposed wiring should be run to simulate normal wiring.

A typical arrangement for table-top equipment is shown in Figure 7.12.

The EUT should be operated within its intended climatic conditions. The temperature and relative humidity should be recorded in the test report.

Equipment intended for table-top use should be stood on a non-conducting table 0.8 m high within the absorber lined screened enclosure. Floor standing equipment should also be stood on a 0.8 m high non-conducting platform where possible. If the equipment is too large or heavy then it may be stood on a 0.1 m high non-conducting support.

The antenna to EUT distance should allow the entire EUT to be in the beamwidth of the transmitted field. However for large EUTs where the face of the EUT is larger than the area of uniformity of the field, several antenna positions will be required to fully illuminate the EUT in a series of tests. For each antenna position a field calibration will be required.

The test should be performed with the antenna facing each of the four sides of the EUT in succession. To minimise testing time a turntable to rotate the EUT should be used if one is available. Tests should be performed for both horizontal and vertical polarisation of each antenna. When equipment can be used in different orientations tests should be carried out on all six sides. This should be by agreement between the manufacturer and user.

The EUT should be fully exercised during testing. This may involve the use of associated equipment or special exercising equipment and may involve preparation of suitable software programs. Every attempt should be made to test under all conditions of operation.

The frequency range should be swept from 26 MHz [80 MHz ENV 50140[21]] to 1000 MHz, using the power levels established during calibration and with the signal 80% amplitude modulated with a 1 kHz sinewave. The sweep may be paused to adjust the RF signal level, or change signal oscillators and antennas. The sweep rate should not exceed 1.5×10^{-3} decades/s. Where the frequency range is swept incrementally, the step size should be less than 10 kHz, 26 [80 MHz ENV 50140[21]] -200 MHz or 20 kHz, 200-1000 MHz, and the dwell time should be greater than 0.1 s. It may be necessary to use a slower sweep rate to ensure that the EUT is fully exercised at each frequency. Frequencies at which the EUT appears to be sensitive may be analysed separately.

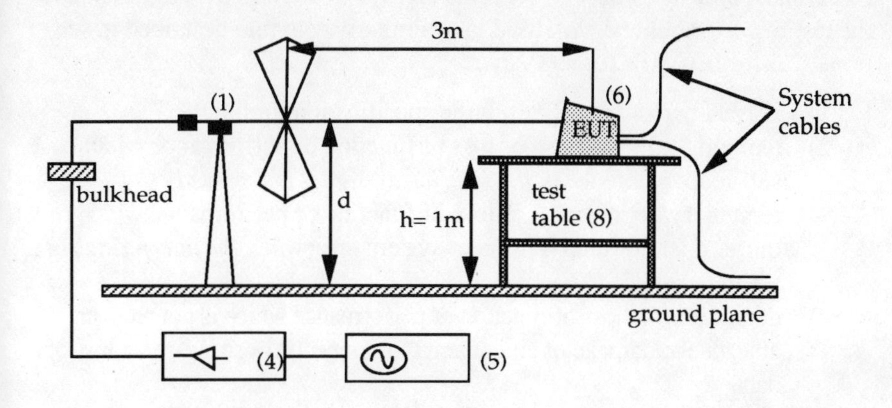

Key
(1) Transmit antenna
(4) Amplifier
(5) Signal generator
(6) Equipment under test
(8) Wooden table (not used for floor standing equipment)
d = height of antenna centre line from groundplane,
d will vary from 1 m to 2 m
h = height of base of EUT above floor, floor standing EUT shall be raised less than 10 cm.

Figure 7.12 ***Typical radiated immunity test set-up within a screened enclosure***

Before performing testing a test plan should be prepared, this should include:

- the size of the EUT
- representative operating conditions of the EUT
- whether the EUT should be tested as table-top or floor-standing, or a combination of the two (for example a system)
- the type of test facility to be used and the position of the antennas
- the type of antennas
- the rate of sweep of frequency
- the severity level to be applied
- the type(s) of interconnecting cables
- the acceptable performance criteria

Some pre-testing may be required to establish aspects of the test plan. The test documentation should include the test conditions (equipment and cable configuration, *etc*), a statement of calibration and the test results.

Evaluation of test results — The draft IEC 801-3 revision[14] suggests that the test results should be classified in a similar way to that described in section 7.2.3 for IEC 801-2: 1991[13], *viz*:

1) normal performance within the specification limits
2) temporary degradation or loss of function or performance which is self-recoverable *when the interfering signal is removed*
3) temporary degradation or loss of function or performance which re quires operator intervention or system reset *when the interfering sig nal is removed*
4) degradation or loss of function or performance which is not recoverable, due to damage of equipment (components) or software, or loss of data.

In the case of acceptance tests, the test programme and the interpretation of the test results are subject to agreement between manufacturer and user.

Implications — There are a number of implications for manufacturers and test houses who are likely to be using the revised version of IEC 801-3, assuming there are no major changes between the draft standard and the final published version:

• the use of a screened room or semi-anechoic chamber is required in order to perform the tests. Unlike earlier versions of IEC 801-3[6] which required a distance of 1 metre between the EUT and the illuminating antenna, the preferred distance is now 3 metres! This means that the minimum working volume, assuming a cubic metre is allowed for the EUT, is now 6 m x 3 m x 3 m (using prEN 55 101-3[10] as a guide). For existing test laboratories, larger screened rooms or semi-anechoic chambers may be required and higher power RF amplifiers may also be required in order to produce the necessary field strength. This represents significant capital investment.

• the field strength uniformity of 0 to +6 dB is likely to be difficult to achieve other than in a semi-anechoic chamber and so smaller manufacturers will tend to be precluded from owning their own facilities.

• partial illumination is permitted for large equipment.

• the EUT operating conditions must be recorded to enable testing to be replicated. The equipment and cable layout should follow normal installation practice as closely as practical.

• in some instances performance degradation is subjective and therefore open to interpretation. It is in the manufacturer's best interests to ensure that full and accurate observations relating to performance degradation are documented in the test report.

7.4 IEC 801 Electromagnetic Compatibility for Industrial-Process Measurement and Control Equipment Part 4: Electrical Fast Transient/ Burst Requirements

General — Experience has shown that industrial instrumentation is vulnerable to fast transients with a high repetition frequency. The object of this standard is to establish a common reproducible basis for evaluating equipment when it is subjected to repetitive fast transients (bursts) on supply, signal, or control lines. The type of transient specified simulates, for example, the interference produced by switching transients; these are produced by switching inductive loads or relay contact bounce.

Test instrumentation — The characteristics of the test generator or Electrical Fast Transient/Burst (EFT/B) generator are specified in clause 6.1 of IEC 801-4[6] and a simplified diagram is shown in Figure 7.13.

The main elements are a high voltage dc power supply which charges an energy storage capacitor *via* a charging resistor. When the energy storage capacitor voltage reaches a particular level the spark gap will 'breakdown', the capacitor will discharge and the pulse waveform will be shaped by the output 'filter'. The resultant transient waveshape is shown in Figure 7.14 when the output of the EFT/B generator is connected to a 50 ohm load.

The operation of the charging circuit or the spark gap (more usually a MOSFET in recent designs), is controlled to produce 15 ms bursts of transients at 300 ms intervals, Figure 7.13. The repetition rates of the transients are also controlled and are defined for peak output voltage values. Pulses of either positive or negative polarity can be produced.

Verification of the characteristics of the EFT/B generator is described in clause 6.1.2 of IEC 801-4[6]. The output is connected to an oscilloscope with at least a 400 MHz bandwidth *via* a 50 ohm co-axial attenuator and the risetime, impulse duration and repetition rate within one burst monitored. Essentially the waveshape should be as defined by Figure 7.14, the pulse risetime is 5 ns +/- 30% and the duration (50% value) is 50 ns +/- 30%. A tolerance of +/- 20% is specified for the repetition rate, which is specified for peak values of output voltage as follows:

0.125 kV	5 kHz
0.25 kV	5 kHz
0.5 kV	5 kHz
1.0 kV	5 kHz
2.0 kV	2.5 kHz

Rc
spark gap
Rm
Cd
U
Cc
Rs
Vo

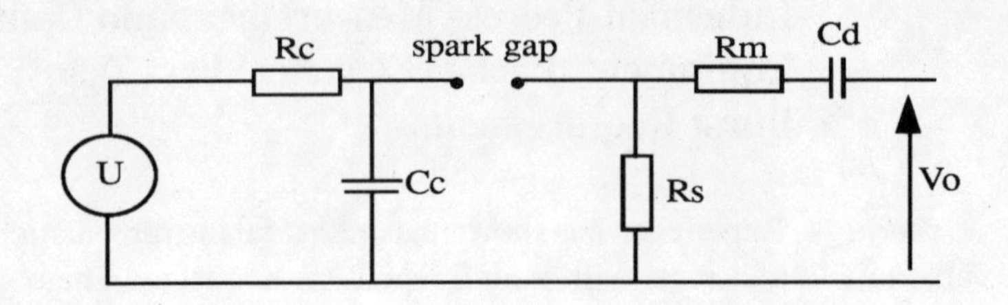

U- high voltage source
Rc - charging resistor
Cc - energy storage capacitor
Rs - pulse duration shaping resistor
Rm - impedance matching resistor
Cd - dc blocking capacitor

Figure 7.13 ***Simplified schematic diagram of the fast transient/ burst generator***

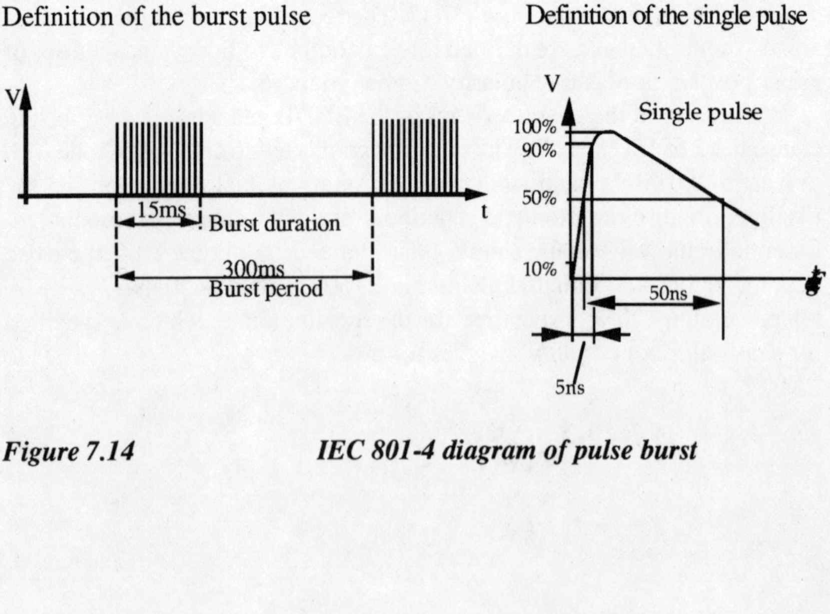

Definition of the burst pulse
Definition of the single pulse
V
15ms
Burst duration
300ms
Burst period
t
V
Single pulse
100%
90%
50%
10%
50ns
5ns

Figure 7.14 ***IEC 801-4 diagram of pulse burst***

It is necessary to couple the generator output to the EUT and two methods are specified:

• coupling/decoupling network for ac/dc supply, Figure 7.15. The coupling section allows the EFT/B generator to be connected symmetrically or assymetrically to the power supply lines via 33 nF capacitors. The decoupling section prevents the pulses from being conducted back into the supply system.

• the second method is to use a capacitive clamp to couple the pulses to signal or control cables having diameters of 4 mm to 40 mm. The clamp consists of two plates 1050 x 140 mm separated by 100 mm high insulating supports. To the top plate hinged along its length are two further plates which are hinged together, see Figure 7.16. The control or signal lines are laid under the hinged plate and rest on the top plate, the clamp should be closed as much as possible to provide the maximum coupling. The generator output is connected between the top and bottom plates and to that end of the clamp nearest the EUT. The clamp itself is placed on a 1 m^2 minimum ground plane which extends beyond the clamp by at least 0.1 m on all sides.

Test severity levels — The test severity levels are shown in Table 7.5, the voltage is the peak voltage of the pulse measured on open circuit with a tolerance of +/- 10%.

Note 'x' is an open level, subject to negotiation between manufacturer and user or specified by the manufacturer. The open-circuit voltages are equal to the energy storage capacitor voltages and will be displayed on the EFT/B generator.

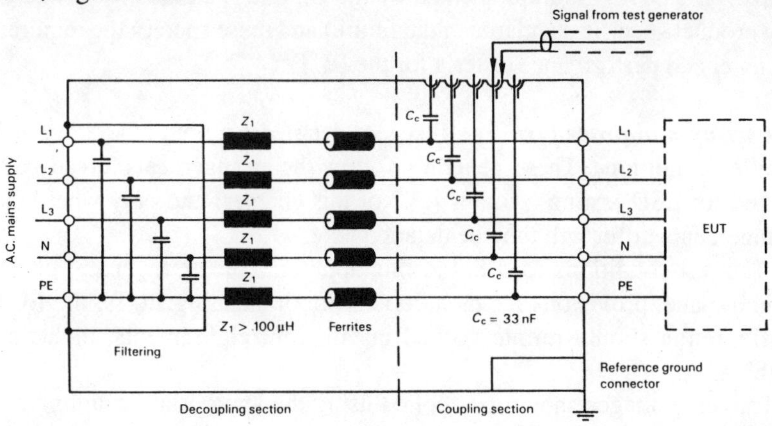

(Example. – Construction for 3-phase lines. D.C. lines/terminals shall be treated in a similar way.)

Warning: The construction and application of the coupling/decoupling network shall be such that existing national safety regulations will not be violated.

Figure 7.15 ***Coupling/decoupling network for ac/dc power mains supply lines/terminals*** © IEC reproduced from 801-4(1) with permission

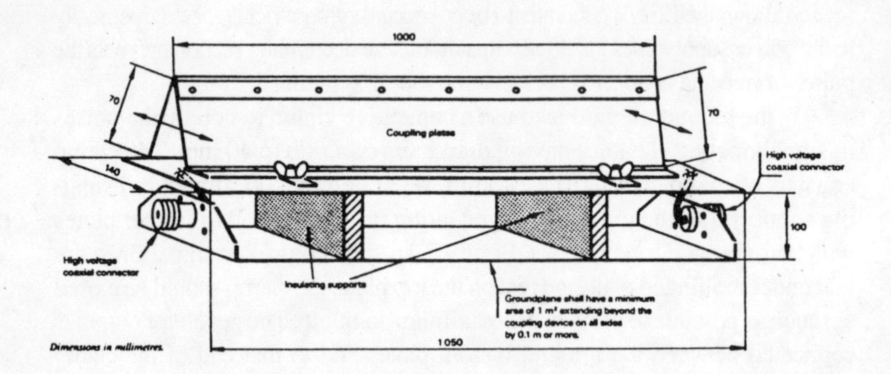

Figure 7.16 ***Construction of the capacitive coupling clamp***

Appendix A of IEC 801-4[6] provides guidance on the selection of test severity level. This is of interest only, since in the context of the EMC Directive, IEC 801-4[6] is implemented by the generic standards (or conceivably product specific standards in the future) and these specify the required test level and performance criteria for the EUT.

Test set-up and procedure — Tests are detailed for both laboratory and installed conditions. These generally follow the arrangements already described for ESD testing (section 7.2.2 of this chapter) and only where the arrangements differ will they be detailed here.

The test conditions for ESD testing apply, *ie.*, the arrangement of the ground plane, provisions for table-top and floor standing EUTs; the EUT configuration should mimic normal installation requirements, including earthing.

The test voltages should be applied using the appropriate coupling device for supply or signal lines. When using the coupling clamp the minimum distance between the coupling plates and all other 'conductive structures' should be 0.5 m, excepting the ground plane beneath the clamp. The length of the signal and power lines between the coupling device and the EUT shall be 1 m or less.

Table 7.5 IEC 801-4 test severity levels

Open circuit output test voltage ± 10%

Level	On power supply	On I/O (Input/Output) signal, data & control lines
1	0.5 kV	0.25 kV
2	1 kV	0.5 kV
3	2 kV	1 kV
4	4 kV	2 kV
x	Special	Special

These open-circuit output voltages, which are equal to the energy storage capacitor voltages, will be displayed on the EFT/B generator.

If the line current is higher than the specified rating of the coupling network, then the test should be performed as for a field test. For these tests the output of the EFT/B generator is connected to each of the power supply terminals and the reference ground. The reference ground is provided by a ground plane of approximately 1 m x 1 m mounted as near to the EUT as possible and connected to the protective earth, the EFT/B generator is positioned on this ground plane and the length of cable between the generator and EUT should not exceed 1 m. If necessary a 33 nF blocking capacitor should be used to connect the generator to the appropriate line [Figure 10 of IEC 801-4[6]]. Note, with these tests no decoupling network is used and care must be exercised to ensure there is no disruption to other equipment operating in the vicinity.

For field tests on signal lines the clamp should be used where possible. However where this is impractical the standard suggests the use of metal foil wrapped around the lines under test, or the generator may be coupled to the lines *via* 100 pF capacitors.

As with the other IEC 801 tests already described the climatic conditions are specified: ambient temperature 15 to 20°C, relative humidity 45 to 75% and atmospheric pressure 68 to 106 kPa.

A test plan should be prepared, which should specify:

- test voltage level
- polarity of test voltage (both polarities are mandatory)
- duration of test > 1min
- number of applications of test voltage
- circuits, lines *etc* to be tested and hence coupling devices required
- representative operating conditions of EUT, including any simulated signal sources
- sequence of application of test voltages

The test plan should be agreed between manufacturer and user and the test level should not exceed the manufacturer's specification, otherwise damage may result.

Evaluation of the test results — The criteria for evaluation of results is broadly the same as for the other IEC 801 methods already considered, the classification of the test results is suggested based on the following criteria:

1) normal performance within the specification limits
2) temporary degradation or loss of function or performance which is self-recoverable
3) temporary degradation or loss of function or performance which requires operator intervention or system reset
4) degradation or loss of function or performance which is not recoverable, due to damage of equipment (components).

The standard also states 'in the case of acceptance tests, the test programme and the interpretation of the test results are subject to agreement between manufacturer and user'.

If IEC 801-4 is being used as a test method under the generic standards then the performance criteria laid down in the generic standard will be used.

7.5 Generic Immunity Standards

7.5.1 BS EN 50 082-1 : 1992 Electromagnetic compatibility — generic immunity standard Part 1. Residential, commercial and light industry.

This generic standard will cover a large proportion of the products falling within the scope of the EMC Directive. The definition of the 'residential, commercial and light industry' location has been described in Chapter 6 for the generic emission standard EN 50 081-1[17].

The main problem with any immunity standard is defining an acceptable level of performance degradation. This is acknowledged in EN50 082-1[9]. It states that 'apparatus shall not become dangerous or unsafe as a result of the application of the tests defined in this standard'.

Certain performance 'criteria' are laid down as follows:

Performance Criteria A: No noticeable loss of function or performance is allowed below a performance level specified by the manufacturer.

Performance Criteria B: No degradation of performance or loss of function is allowed below a performance level specified by the manufacturer. During the test, degradation of performance is allowed. No change of actual operating state or stored data is allowed.

Performance Criteria C: Temporary loss of function is allowed, provided the function is self recoverable or can be restored by the operation of the controls

The standard specifies that the manufacturer must provide a definition of performance criteria, during or as a consequence of the EMC testing and this should be included in the test report.

The tests are defined in four tables and are applicable to the ports defined in Figure 6.6. The tests are to be carried out where the relevant ports exist. Further the standard states that the when the electrical characteristics and usage of the apparatus are considered 'it may be determined...that some of the tests are inappropriate and therefore unnecessary'. Should such a decision be taken then this should be recorded in the test report.

Measurement methods and limits are defined using a series of tables which call up IEC 801[5] parts 2, 3, and 4. The draft of this standard also included IEC 801[5] parts 5 and 6 which are not yet agreed, and also made reference to other unpublished documents. These are now included in the informative annex to the standard and it is to be expected that when IEC 801[5] Parts 5 and 6 (or their IEC 1000 equivalents) are agreed the generic standard will be revised. The operating conditions during testing are as defined in the generic emission standard *ie* as EN 55 022[18], excepting that the configuration/position should be in the most susceptible operating mode for the frequency band being investigated, consistent with normal applications.

The specified immunity test levels are:

i) an unmodulated RF field strength of 3 V/metre, over the fre quency range 27-500 MHz [IEC 801-3: 1984[6]]
ii) an electrostatic discharge of 8 kV applied by the air discharge method [IEC 801-2: 1984[6]]
iii) and fast transient bursts applied through a capacitive network of 500 V applied to signal lines and 1 kV to power lines [IEC 801- 4: 1988[6]].

An informative 'Annex A' is included which includes a whole host of additional tests which it is proposed will be included in the standard when the 'basic' standards are agreed. These are either under consideration, only exist as IEC committee documents or are provisional standards. An example is prEN 50 093[19] which specifies voltage dips as follows:

the voltage variations on the mains supply are defined as a 30% dip for 10 ms a 50% dip for 100 ms and a 95% interruption for 5 seconds. It should be noted that only performance criteria 'C' apply to the 50 and 95% dips.

7.5.2 BS EN 50082-2: 1995 Electromagnetic Compatibility - generic immunity standard Part 2: Industrial

The generic immunity standard for the 'industrial' environment was finally published in mid 1995 having been previously circulated for public comment in 1991 [91/21828 DC]. Elsewhere in this book it is referred to as 'prEN 50 082 - 2' its status when this second edition was originally published.

The definition of the industrial environment and equipment ports is the same as for the generic industrial emission standard [EN 50 082 - 2[20]] described in 6.4.1.2. The reference standards are the 'basic' EN 61000 - 4 series of standards and ENV standards 50140, 50141 and 50204. Generally the industrial generic immunity standard is similar to EN 50 082 -1[9] and calls up the reference standards on a port by port basis using tables, but with higher severity levels of testing reflecting the more arduous environment.

For the 'enclosure' port radiated immunity testing is specified to be performed as described in ENV 50140[21]. The severity level specified is 10 v/m over the frequency range 80 MHz to 1 GHz and 80% amplitude modulated with a 1kHz tone. In addition a spot frequency test at 900MHz +/- 5MHz of 10v/m is specified to be pulse modulated with a 50% duty cycle and 200Hz repetition frequency, as ENV 50204 and the test set up of ENV 50140. A power magnetic field test is also specified of 30 A(rms)/m, 50 Hz, the test method being described by ENV 61000-4-8, where the EUT is considered to be susceptible to magnetic fields, *eg*. it contains Hall elements or electro dynamic microphones. It is noted that CRT display performance may degrade above 3 A/m! Criterion A is specified for these tests. ESD testing is specified for the enclosure port with levels of 4 kV and 8 kV for contact and air discharge respectively, as EN 61000-4-2 (identical to IEC 801-2:1991[13]), criterion B.

Tables are included for conducted immunity tests applied to: signal lines, data buses, but not involved in process control; process, measurement and control lines, and long bus and control lines; ac and dc power ports; and the earth port. The tests specified are: 150 kHz - 80 MHz rf carrier of 10v, 80% amplitude modulated with a 1 kHz tone (150 ohm source impedance) coupled common mode to the ports (cables) as specified by ENV 50141; and 'electrical fast transient/bursts' coupled as specified by EN 61000-4-4 (IEC 801-4:1988[6]) at levels of 1 or 2 kV.

A detailed examination of the standard should be made to establish the test plan for the EUT, taking account of the various notes and performance criteria.

7.6 Summary of Implications

In this chapter immunity standards have been reviewed and a number of implications highlighted. Standards considered have been:

'Basic' standards

- IEC 801-2: 1984[6] and IEC 801-2: 1991[13] (also prEN 55 101-2[10]) covering ESD
- IEC 801-3: 1984[6] and 90/29283[14] the draft revision to IEC 801-3 (also prEN 55 101-3[10]), and ENV 50140[21] for RF fields.
- IEC 801-4: 1988[6], the application of conducted electrical fast transient/bursts

Generic standards

- EN 50 082-1[9] the generic immunity standard for 'residential, commercial and light industry' locations
- prEN 50 082-2[11] the draft generic immunity standard for the 'industrial' environment

These standards have been used to illustrate a number of points, principally:

- the need to have access to suitable test facilities. This may involve companies in significant capital investment or the selection of a third party facility. In either case additional costs will be incurred.
- unlike many other types of electrical equipment testing, there is considerable uncertainty in the measurements made. Repeatability of testing is also a concern. Reasons for this are that:
  i) equipment configuration is open to interpretation
  ii) the calibration method used for test facilities need further research
- the generic standards apply to an environment type, and therefore cover a broad range of products. For measurement methods they make reference to published standards.

These conclusions concur with those to Chapter 6 for emission standards, in addition however, the following have been demonstrated:

- the degree of performance degradation may be subjective and in many instances the onus will be on the manufacturer to select a suitable test severity level for his equipment
- the revised standards for ESD testing, whilst offering an improved technical solution, are likely to increase the test duration and hence there will be a cost penalty
- for radiated immunity testing, limitations on the test methods have been identified in respect of screened rooms, the stripline circuit and TEM cells. Particularly in the case of the stripline, IEC 801-3: 1984[6] the standard has been shown to have technical deficiencies. This is significant as this standard is called up by the generic standards
- in practice it has been shown that equipment tested in accordance with IEC 801-3: 1984[6] can still suffer maloperation when installed in a given environment. Manufacturers are required to comply with the essential protection requirements of the EMC Directive and not with standards. Manufacturers must therefore use the standards with caution and with careful consideration of the application. Also there is a requirement on the stand-

ards bodies to be continually ensuring that test methods represent as far as possible the conditions likely to be found in practice

• compared with IEC 801-3: 1984[6] the proposed revision to IEC 801-3[14] and ENV 50140[21] have implications for existing users of EMC test facilities:

i) the increase in measurement distance between the EUT and antenna from 1 m to 3 m means that the minimum working space within a screened enclosure or anechoic chamber is likely to be 3 m x 3 m x 6 m; a signifi cant increase in size

ii) in order to achieve the specified field strengths higher power RF am plifiers will be required

iii) the field uniformity has been defined, for a uniform area, without the EUT present rather than measuring the field strength adjacent to the equipment. This field uniformity may however prove difficult to achieve in all but the largest anechoic or semi-anechoic chambers and more re search into the acceptable field variation is required. This is partly miti gated by the change in frequency range (80 MHz - 1 GHz) defined in ENV 50140[21]

• the generic standards allow a manufacturer flexibility to apply all or some of the tests specified, whilst this may be perfectly sensible it also allows for a variety of interpretation, such that for the same type of product it is possible that different tests may be applied by different manufacturers. This delegation of responsibility to a manufacturer may actually be regarded as consistent with the self-certification route to compliance with the EMC Directive, as the manufacturer is required to make his own declaration of conformity with the protection requirements of the Directive.

References

1. 89/336/EEC Council Directive 'on the approximation of laws of Member States relating to electromagnetic compatibility', Official J. of the Eur Comm. No.139 25 May 1989, pp 19-26
2. Mil-Std-461 'Military Standard Electromagnetic Emission and Susceptibility requirements for the Control of Electromagnetic Interference, Department of Defense, USA
3. Def Std 59/41 'Defence Standard 59-41 Electromagnetic Compatibility', Min. of Defence, UK
4. ANSI C63.2-1987, Standard for Instrumentation - Electromagnetic noise and field strength, 10kHz to 40 GHz - Specification, American National Standards Institute
5. SAMA standard PMC 33.1 - 1978 'Electromagnetic Susceptibility of Process Control Instrumentation', Scientific Apparatus Makers Association
6. IEC 801 Electromagnetic Compatibility for industrial-process measurement and control equipment, Part 1: 1984 General introduction; Part 2: 1984, ESD; Part 3: 1984 Method of evaluating susceptibility to radiated electromagnetic energy (BS 6667: 1985 as IEC 801 Parts 1, 2 and 3); Part 4: 1988, Electrical fast transient/burst requirements

7. HD 481 CENELEC implementation of IEC 801 Parts 1, 2 and 3
8. CEGB 'Radio frequency interference susceptibility testing of electronic equipment', Specification CEGB- DN 5 Issue 2 April 1987, Central Electricity Generating Board
9. BS EN 50 082-1: 1992 'Electromagnetic compatibility - generic immunity standard Part 1. Residential, commercial and light industry', British Standards Institution, 1992
10. prEN 55 101 Immunity of Information Technology Equipment; Part 2 (BSI 89/34172 DC) ESD; Part 3 (BSI 89/34171 DC) immunity to radiated electromagnetic energy; Part 4 (BSI 90/30270 DC) immunity against conducted signals; NB.to be renumbered prEN 55 024 Parts 2, 3 and 6
11. prEN 50 082-2: 1991 Generic immunity standard, Industrial (BSI 91/21828 DC)
12. B Jones 'Improvement to the ESD Testing of Equipment', IEE EMC Conf., York, 1990
13. IEC 801-2: 1991 'Electromagnetic compatibility for industrial process measurement and control equipment Part 2: Electrostatic Discharge', IEC, 1991
14. 90/29283 DC Draft - revision of IEC publication 801-3: electromagnetic compatibility for industrial-process measurement and control equipment Part 3: immunity to radiated radio frequency electromagnetic fields, British Standards Institution, 1990
15. IEC Guide 107: 1989 'Electromagnetic compatibility. Guide to drafting of electromagnetic compatibility publications'
16. Dawson, Mann and Marvin, 'Improving repeatability of immunity measurements within screened enclosures', IEE EMC Conference, Herriot-Watt September 1992
17. BS EN 50 081 part 1 'Electromagnetic compatibility - generic emission standard generic standard class: residential, commercial and light industry', British Standards Institution, 1991
18. EN 55 022: 1987 (BS6527:1988) 'Limits and methods of measurement of radio interference characteristics of information technology equipment', British Standards Institution, 1988
19. prEN 50 093: 1991 'Basic immunity standard for voltage dips, short interruptions and voltage variations'
20. 91/21829 Draft British Standard BS EN 50 081-2 (prEN 50 081-2) 'Electromagnetic compatibility - generic emission standard generic standard class: industrial', British Standards Institution, 1991
21. DD ENV 50140: 1994 'Electromagnetic compatibility - Basic immunity standard - Radiated radio-frequency electromagnetic field - immunity test', British Standards Institution, ISBN 0 580 22607 7

8
Test Facilities

This chapter is concerned with identifying the major types of test facility required for performing EMC testing in accordance with the standards. Different interpretations of test facility are illustrated and the relative cost implications discussed. Therefore this chapter should be read by those tasked with establishing or considering establishing EMC testing facilities. The requirements of Open Field Test Sites and their calibration are described, along with practical OFTS implementations (examples are given of low cost and all weather sites). Screened rooms are discussed, in particular their limitations in respect of reflections and resonances. Anechoic chambers are also considered, particularly the effects of absorber depth and the cost implications. Hence the reader is presented with the simple facts, including cost considerations, involved in the choice of EMC test facility types. These may affect investment plans or choice of a test house.

8.1 Introduction

In June 1989 the Department of Trade and Industry (DTI) published the report prepared for them by the consultants W S Atkins[1]. This report discussed the provision of EMC test facilities within the United Kingdom (UK) and the potential demand for these services, based on estimates of the number of products which would require testing after 1992. The essential conclusion of this report was that the mismatch between the test facilities available and the number of test facilities required and was of the order 6:1. When the provision of Open Field Test sites was considered, the mismatch was found to be of the order 12:1.

The establishment of EMC test facilities requires considerable capital investment. The difficulties faced by industry and entrepreneurs alike have been the number of uncertainties surrounding the EMC Directive, particularly the requirements, or potential requirements, of the harmonised standards, interpretation of the scope of the Directive [Imeson 1990[2]] and the effect of the delay of full implementation until 1 January 1996. Many of these uncertainties were resolved with the adoption of the amending Directive, publication of the explanatory document and the first of the generic standards. Assuming the worst effects of economic recession are past, then investment in further EMC test facilities will accompany the increase in business confidence.

This chapter describes the main types of EMC test facility and discusses their advantages and disadvantages. In particular, open field test sites are considered and their practical implementations illustrated to demonstrate differing approaches to construction.

8.2 Open Field Test Sites

An Open Field Test Site (OFTS) or Open Area Test Site (OATS) is required for measuring radiated emissions from equipment. The essential requirements are laid down in CISPR 22[3]. Practical sites have been in existence for a number of years and used by industries required to comply with the US FCC regulations, 47CFR[4] Part 15 and Part 18 (see Chapter 12) and the German VDE standard 0871[5] (also described in Chapter 12). Implementation of the EMC Directive will effectively enforce EN 55 022[6] and the Generic emission standards (Chapter 6). The requirements for an EN 55 022[6] site are identical to CISPR 22 and are described in Chapter 6 section 6.1. The only official guide to constructional details is given in the FCC document OST55[7].

The essential features described in EN 55 022[6] and OST55[7] are summarised below:

1) The test site shall be flat, free of overhead wires and nearby reflecting structures and its area is defined by Fig 1 p14 of EN 55 022[6] (Figure 6.2). The equipment under test (EUT) and measurement antenna form the focii of an ellipse where the major diameter is twice the distance 'd' between the EUT and antenna and the minor diameter $\sqrt{3}\,d$. The distance between the source and measuring antenna is usually 3, 10, or 30 m depending on the specification in use and the physical size of the equipment under test (*see* Section 6.1.3 and Figure 6.3). OST55[7] states that any object greater than 5 cm in dimension is to be cleared from the surface of the site.

2) Objects outside the elliptical boundary, such as buildings, parked 'automobiles' may still affect the measurements — hence care should be taken

to choose a location as far as possible from large objects or metallic objects of any sort [OST55[7]].

3) A recommended minimum size for a metallic ground plane is given in Fig 3 p15 of EN 55 022[6] (Figure 6.3). This is usually simplified to a rectangle having a width of twice the maximum test unit dimension and extending 1 m beyond the perimeter of the test unit and 1 m beyond the measurement antenna. The ground plane should have no gaps or voids that are a significant fraction of a wavelength at 1 GHz. The recommended mesh size for perforated metal ground planes is 1/10 of a wavelength at 1 GHz — about 30 mm [EN 55 022[6]]. OST55[7] recommends a mesh of either 1/4" or 1/2", it also recommends that the metal ground plane should be the size of the ellipse, particularly if the site is a paved area or a rooftop installation. 'Acceptable performance may be achieved without a metal ground plane, provided that the ground plane is composed of homogeneous good soil (not sand or rock) free of buried metal' [OST55[7]].

4) OST55[7] recommends that the measuring equipment and test personnel should be in the plane of the antenna orthogonal to the site axis and at least 3 m from the antenna.

5) OST55[7] provides notes on weather protection enclosures, particular attention is drawn to limiting the dimensions of metal objects above the ground plane to less than 3 cm.

The accepted measure of site calibration is 'site attenuation'. This is defined as the ratio of the source antenna input power to the power induced in a load connected to the receiving antenna. Methods for site attenuation measurement are specified by OST55[7] and VDE0877[8]. An antenna transmitting a known power replaces the EUT and the received power is measured. Cable and instrumentation errors are corrected by repeating the measurement with the two antenna feeders connected directly together. Typically the measured site attenuation must be within +/-3 dB of the calculated response. Tuned dipoles are usually used for site attenuation measurements. However, experience has shown that, particularly at 3 m, the two antennas couple to each other very well and the site has very little effect on the results. This can hide imperfections of the site which may affect the measured results from an EUT where the source and receiving antenna are not so well coupled. Other inaccuracies in site attenuation measurements are influenced by the assumptions made when calculating the response, *viz*:

1) The measurements are dependent on the antenna factors, these are

either a manufacturer's data or have been measured by a standards laboratory, and assume far field radiation patterns. However, at 3 m and at lower frequencies (50 MHz and below) there will be some degree of near field coupling.

2) The antenna factors assume conditions of free space which is not the case when a ground plane is present.

3) Reflections from the ground plane will also introduce errors into the measured field when it is compared with the calculated value. The effect will vary with frequency as the positions of field maxima are dependent on the distance between the source and receiving antennas and the height of the source antenna. The maximum received signal is located by scanning the receiving antenna over a specified height to determine where the direct and reflected waves add in phase.

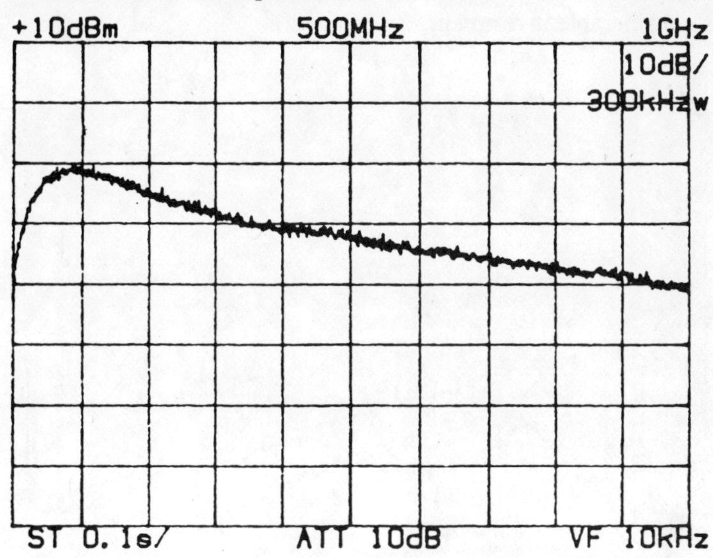

Figure 8.1 ***Output power into 50 ohms***

An alternative method of calibrating an open field test site is to use the Comparison Noise Emitter or CNE. This device was originally devised as a research tool for characterising screened rooms and is a 30 MHz to 1 GHz broadband noise source, Figure 8.1. The response of the CNE on an OFTS is shown in Figure 8.2. Comb generators have previously been used to compare the characteristics of different open sites, Bronaugh[9]. These suffer from the disadvantage that the generated spectrum is not continuous and therefore discrepancies due to resonances or antenna characteristics may not be identified. NAMAS use the CNE to compare the performance of all the UK accredited sites. An improved version, the MK III, with a frequency range from a few kHz, a slightly higher output power and a flatter response is now available from the York Electronics Centre.

The use of the open field test site is assured in the immediate future by the requirement placed on industry to comply with the EMC Directive and therefore with EN 55 022[6] and EN 55 011[10], either directly or by virtue of the generic standards. Unfortunately environmental considerations may preclude the use of such a site. For example a 30 m test site will require an area of 60 m x 52 m and such an area is likely to be prohibitively expensive in a densely populated industrial area. Also in an industrial or urban area the ambient electromagnetic environment is likely to be too high to enable meaningful measurements to be made (see Chapter 6 for ambient requirements). In the UK the climate presents practical difficulties when the viability of an outdoor open site is considered and to cover a site with a building transparent to EM waves requires considerable investment.

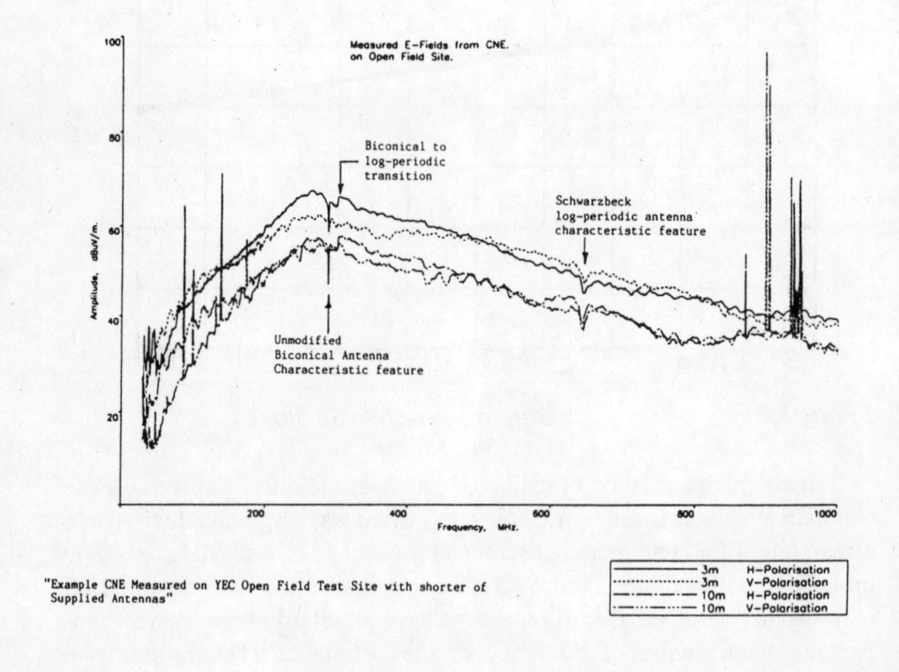

Figure 8.2 Response of the Comparison Noise Emitter on an OFTS

8.3 Practical Implementations of Open Field Test Sites.

8.3.1 Outdoor 10 m site - YEC, University of York

Figure 8.3 shows the site plan for the York Electronics Centre's OFTS. When its features are compared with those listed as ideal in section 8.2, a number of discrepancies are apparent:

1) the proximity of buildings and the car park.
2) the openings in the ground plane mesh exceed 30 mm in one direction.
3) a metal door lock and handle are fitted to the glass reinforced polyester (GRP) building used to protect the equipment under test (EUT) from the elements.

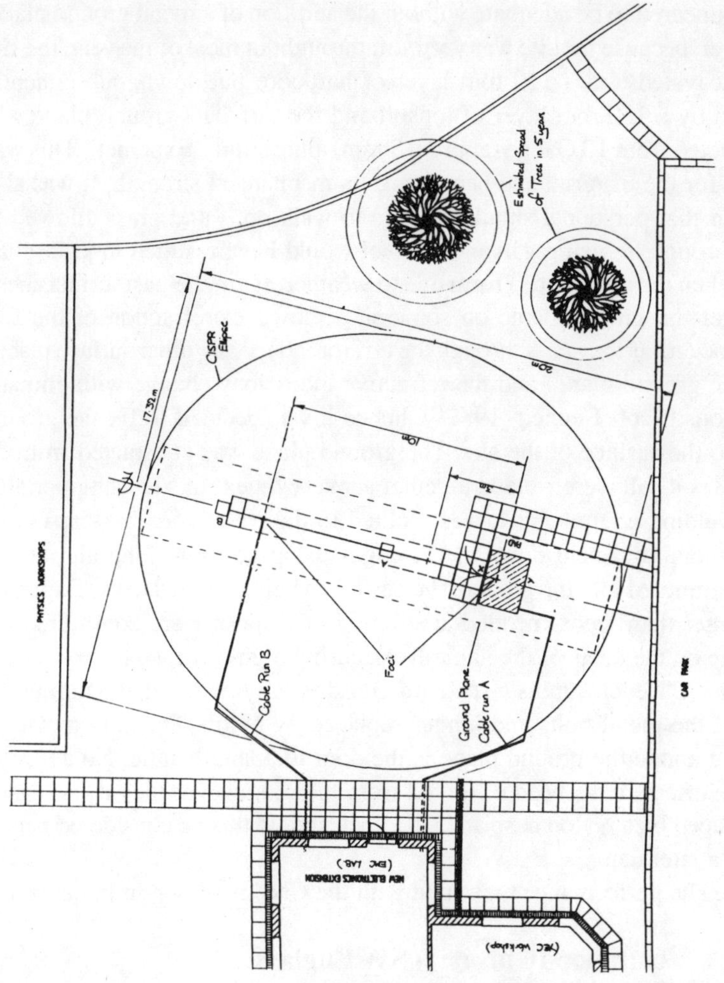

Figure 8.3 ***Site plan of York Electronics Centre's OFTS***

The discrepancies listed above were all recognised at the pre-construction stage and were accepted as an 'engineering compromise'. Prior to building the site the ambient was monitored and it was determined that the proposed site was 'quiet', excepting broadcast bands and intermittent mobile transmissions. An approximate site attenuation measurement was also made which indicated that the proposed site did not appear to be adversely affected by the proximity of the buildings or the car park. A non-technical factor which weighed heavily in favour of the location was the security aspect, a more remote location on the University campus might well have been subject to vandalism in what is effectively a public place.

At this location the water table is particularly high and the natural ground plane appeared to be adequate without the addition of a metal ground plane. However, because the site was very soft throughout most of the year, the site was excavated and a 150 mm layer of hardcore put down, subsequently covered by a 150 mm layer of topsoil and the turf. The ground plane was constructed from 10 Gauge (approx. 3 mm) aluminium 'Expamet'. This was chosen for the material and because of its mechanical strength. It was also required that personnel should be able to walk on it and grass allowed to grow through it. Using galvanised steel would have resulted in killing the grass when the zinc oxides form due to weathering. In the past turf has been laid over the ground plane on some sites, however refraction of the EM waves occurs as they pass through the turf (or indeed any other surface placed over the ground plane) and the refractive index may change with climatic conditions [Scott Bennett, 1982[11]], hence it was decided to fix the ground plane to the surface of the site. The ground plane was constucted from 24 sheets 8' x 4', all were welded together at every 'apex' *in situ* using portable MIG welding equipment. It was secured to the paved area by screws and 'Rawlplugs' and held down at the edges using 450 mm long aluminium pegs hammered into the ground, the pegs then being welded to the Expamet. The larger than recommended apertures in the mesh were considered acceptable on the basis of the naturally occuring high water table.

The EUT enclosure is a standard Glasdon 'Olympic' GRP building but with all the metal bolts and hinges replaced by nylon. The only metal remaining above the ground plane is the door lock and handle. No apparent adverse effects have been observed from these, however they could easily be replaced by a Nylon hasp and a padlock should this be considered necessary at a later date.

The site performance measured with the CNE is shown in Figure 8.2.

8.3.2 Outdoor 10 m site - SW England

Site attenuation measurements had been carried out on this site using the

VDE method. The results showed that the variation in performance compared to the theoretical to be in excess of +/- 3 dB. The site was inspected in 1990, and an approximate site plan prepared, Figure 8.4.

When a comparison was made between this site and the features outlined in 8.2, the following observations were made:

i) site flatness — the site surface consisted of concrete foundations from earlier structures, tarmac and a hard wearing epoxy surface over the ground plane. A major discontinuity of the order of 100 mm existed all along one side of the ground plane.
ii) some of the concrete foundations showed evidence of sawn off steel structures, there may also have been steel reinforcement in the concrete — this may have been a factor in the reflection properties of the site.
iii) the road through the plant passed through the ellipse and had some camber.
iv) the wooden posts along the edge of the ground plane, although non-conductive, may become conductive in wet weather and affect the site characteristics.
v) the control room was positioned partially within the ellipse. Interpreting the FCC requirements, this should be outside the ellipse and in line with the antenna positioning mast. The antenna cable should also be at right angles to the axis of the site and in the plane of the antenna.
vi) inspection of the site led to the conclusion that the water table was low and the site was unlikely to have a good natural ground plane properties.
vii) the turntable used for rotating the EUT had a gap around its periphery of the order of 1 cm, although bonded to ground, the table was not electrically in contact with the ground plane.
viii) the weatherproof enclosures had aluminium window frames.
ix) beyond the boundary of the ellipse were:
- a trailer park with containers (large metallic objects!)
- a store with a metal door
- a wire netting fence

x) the switchgear positioned behind the EUT enclosure appeared to be within the ellipse and was also above the ground plane.

A number of recommendations were made to the client and it is understood that the site has been reconstructed to take account of as many of the factors described as practical. The reconstructed site was commissioned in 1991.

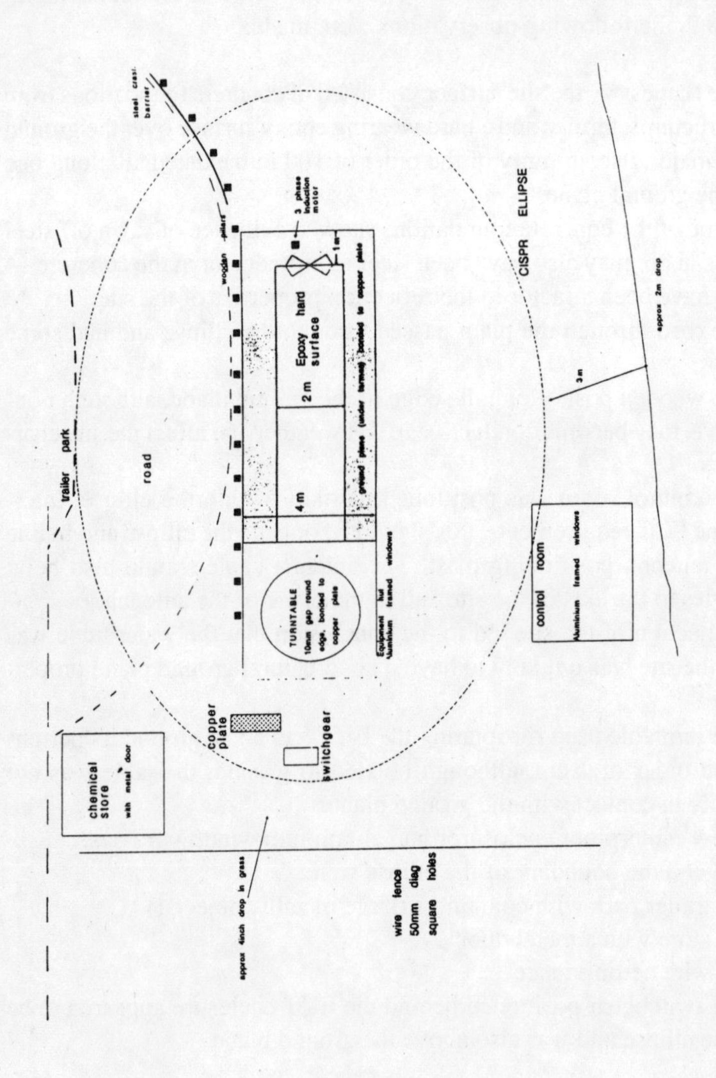

Figure 8.4 ***Site plan outdoor 10 m site SW England***

8.3.3 All weather open field test sites

A number of large companies have invested in all weather open field test sites. For example IBM (UK) have two sites in the UK at Hursley Park and Greenock, Assessment Services have a sites at Titchfield and Bearley near Birmingham, British Telecom have a site at Martlesham. Further sites are known to be under consideration.

These sites are similar in construction whilst differing in detailed design and consist of a ground plane covered by a glass reinforced polyester (GRP) superstructure, with the measurement instrumentation and test personnel situated below the ground plane. Provision for rotating the EUT is provided by means of a turntable. A summary of the main features of the Assessment Services site near Birmingham is given below as a typical illustration.

The construction of this site was described by M.Robinson and A.R.Nash[12] 'Construction of an Open Area Test Site' and published in the EURO EMC 90 conference proceedings, the main features may be summarised as follows:

i) *Environment* — a quiet rural site was chosen meeting the requirements of the standards being less than 34 dbuV/m *ie.*
6 dB below the specified emission limits at 10 m.

ii) *Superstructure* — the requirements for the site covering were that it should be non-conductive, possess thermal insulation and have a smooth exterior to prevent the build-up of dirt and debris which might prove to be conductive to some degree. This was achieved by using a sandwich construction of two GRP skins with a foam filling acting as a binding agent and providing thermal insulation. Colour dyes having metallic substances in their composition were avoided for the GRP and a neutral grey was chosen. The dimensions of the superstructure are 20 m x 12 m x 7 m high, this covers the entire ground plane and allows for an antenna height scan of 4 m when using a tuned dipole. Natural lighting is provided through the north elevation to prevent excessive solar heating which during the summer could render the building uncomfortable and cause excessive expansion of the welded ground plane. Equipment access is provided through double doors at the turntable end of the building.

iii) *Ground plane* — the size chosen was 20 m x 12 m using the criteria already discussed. This was manufactured to tolerance of +/- 5mm. A lattice work of rolled steel channels was bolted down and shimmed to give the required flatness. A structural concrete screed was laid within and flush with the lattice work and finally 6 mm thick galvanised steel sheet was welded to the lattice. Light wells for illumination of the building were incorporated beneath the ground plane and covered with a steel mesh bonded to it.

iv) *Control room* — the measurement equipment and test personnel are situated in a basement beneath the ground plane ensuring that the measured emissions are wholly attributable to the EUT and not the measuring equipment. Special construction techniques have been used to ensure a waterproof basement structure. A feature of this installation is the excavation at one end of the site to allow daylight into the working area.

v) *Turntable* — a 5 m diameter 3 tonne capacity turntable has been incorporated flush with and connected to the ground plane. The peripheral gap between the ground plane and turntable does not exceed 5 mm and contact is maintained using phosphor bronze bristle brushes. Incorporated within the turntable structure and bonded to the ground plane are power filters and data filters to enable the required isolation to be achieved between the EUT and its exercising equipment located in the basement. The turntable is controlled for setting up purposes by a wander lead above the ground plane and is computer controlled from the basement during testing. The turntable has 370 degrees of movement and rotates at two speeds 1 rev/min and 1/4 rev/min.

It is interesting to compare this site with those described in 8.3.1 and 8.3.2.

8.4 Screened Rooms and Anechoic Chambers

In addition to the OFTS the major EMC testing environments are the Screened Room and the Anechoic Chamber. The advantage of the screened room and chamber over the OFTS is that the external ambient environment is excluded. The major disadvantages are:

i) the limitation placed upon the size of EUT which may be tested; this is dependent upon the room or chamber size,
ii) that the room or chamber can become a dominant part of the electro magnetic system due to reflections and resonant modes and
iii) in the case of anechoic chambers the cost may be prohibitive.

Because of the technical disadvantages indicated, screened rooms in particular are of limited use for radiated emission measurements. They are however essential for radiated immunity measurements when large fields are intentionally generated which, unless shielded from the external environment, may present a threat to broadcast services or electrical equipment used locally. For all other types of immunity testing *eg*. conducted or ESD, the environment provided by the screened room is essential to prevent interference with other laboratory equipment.

8.4.1 Screened rooms

A screened room (or Faraday cage) is usually in the form of a large metal box (*eg*. 5 m x 4 m x 3 m) constructed from sections which are bolted together with metal gaskets between the sections. The room shields equipment within it from the external ambient environment or, conversely, prevents radiated electromagnetic energy generated within the room from affecting the external environment. The power supply and signal cables to the EUT are filtered, with the filters being bonded to the skin of the screened room completing the isolation.

When the screened room is compared with an OFTS it is apparent that it takes up relatively little space and is likely to be less expensive. A screened room can also be installed within an existing test facility which may simplify the logistics of EMC testing compared with using an OFTS which may be some distance from other facilities. The screened room is also available for testing independent of the weather conditions. This is not true for the OFTS unless the site is covered with an expensive building transparent to radio waves as already discussed.

Although the climatic and ambient environmental problems associated with an OFTS are eliminated by the screened room, it does have inherent properties associated with its metallic construction and dimensions which present limitations on its usefulness. The screened room behaves as a waveguide cavity resonator having a cut-off frequency below which fields do not propagate efficiently [Harrington[13]] and above which they exhibit a series of high Q resonances [Keiser[14]]. Because several modes of wave propagation may exist at a particular resonant frequency, significant changes in field strength may occur over relatively small distances, particularly for higher order modes. This means that it is very difficult to obtain repeatable measurements on the same piece of equipment, as they are sensitive to both the positions of the antenna and the EUT. Because the resonances are a function of room dimensions it is difficult to obtain correlation between measurements performed in different rooms. Where measurements are made inside a room the uncertainty can be very large, of the order +/- 40 dB [Marvin and Dawson[15]].

A further consideration is that antenna correction factors are usually given for plane wave conditions and are therefore unlikely to yield reliable values for field strength measurements within a screened room since the propagation is unlikely to be plane wave (as already discussed). The proximity of the conducting walls may also capacitively load the measurement antenna, again introducing errors.

It is apparent that relating screened room radiated emission measurements to those made on an OFTS is extremely difficult. It is also difficult to establish a uniform field to illuminate an EUT over typically 27 MHz to 1 GHz when performing radiated immunity testing and to measure the field during the testing. The screened room does however enable the emission frequencies of an EUT to be characterised prior to carrying out measurements on an OFTS where amplitudes can be measured more accurately. This is particularly helpful when emissions are close to broadcast frequencies for example, as they may be hidden by the ambient noise on the OFTS, but may in fact be close to or above the emission limits laid down in the standards. Screened rooms are also useful for performing conducted emission and immunity testing making use of the filtered power supply and minimising the pick-up of external signals by the cables. The large fields which can be generated by conducted immunity test equipment are also prevented from interfering with the local environment.

A number of methods have been devised to improve measurement accuracy these include using a 'hooded antenna', building an underground chamber without metal walls, stirring the modes and damping the cavity resonances. This latter technique was developed at the University of York [Marvin and Dawson[15]], although only a partial solution it has a relatively low cost and is discussed further in Appendix F.

The 'hooded antenna' entails surrounding the antenna with a shielded box lined with absorbent tiles. The box is open in the direction of the EUT only. The intention is to reduce reflections from the conducting walls *etc.*, so that the antenna receives only the direct wave. However, this method does not work well below 200 MHz as the ferrite tile absorbers are not effective at low frequencies and it is below 200 MHz that the greatest problems exist due to depth and cost of absorber required to make the room anechoic (see next section).

The idea behind an underground chamber is that the earthen walls will be lossy and the Q of such a cavity will be low [Donaldson, Free, Robertson and Woody[16]]. ICL commissioned such a chamber which has been converted from a salt mine in Cheshire. In this facility it has been found that additional absorber was required in the roof space. This facility is now operated under the auspices of Design to Distribution Ltd.

Stirred mode chambers are screened rooms with the addition of at least one large rotating paddle which is used to 'stir' the modes in the room. The effect of the paddle is to adjust the resonant frequencies and positions of field maxima so that the average of the radiated power received by the antenna is proportional to the power radiated by the source into the room. It is necessary for the power to be averaged over a period which is greater than the period for one revolution of the paddle at each frequency of interest. This means that measurements may take significantly longer than with other methods. Also stirred mode measurements can only work if every frequency becomes resonant at some point during the rotation of the paddle, hence the technique cannot be used below the cut-off frequency of the room, which is dependent upon its size and therefore cannot be used for low frequencies. Stirred mode chambers are not particularly useful for radiated immunity measurements because, although the EUT may operate satisfactorily in the average field, the peak levels may be significantly higher causing the unit to fail hence the EUT may only appear to operate correctly in an artificially low average field.

8.4.2 Anechoic and semi-anechoic chambers

An anechoic chamber consists of a screened room with its walls, ceiling and floor lined with radio absorbent material (RAM) to reduce reflections of the incident wave. Ideally the chamber models free space with no reflections of the emitted wave and therefore no resonances.

The RAM is usually a carbon loaded foam with a defined conductivity. At microwave frequencies multilayer foam is used enabling the conductivity to increase with the depth of the material and obtaining a gradual modification in wave impedance, so that a plane wave incident upon the surface of the block is absorbed and not reflected. At lower frequencies the same effect is produced using uniformly loaded foam shaped into pyramids. The depth of the foam should exceed 1/4 wavelength to work most effectively, however for EMC testing 1/5 wavelength is generally considered adequate. Absorbers are not as efficient when they are within the near field of the source where, depending on the type of source and the distance from the absorber, the wave impedance is undefined.

The depth of absorber required for anechoic performance means that for frequencies lower than 200 MHz it is very expensive to build a chamber, as the size will be large and therefore a large amount of absorber required. For example, at 30 MHz the pyramids need to be 2.5 m deep, hence a room large enough to accommodate the absorber will be a 5 m cube and have no operating space at all! At 30 MHz for all the absorber to be in the far field of the source an internal dimension of 3.3 m is required. Therefore for a point source the minimum room dimension is now 8.3 m.

An environment approximating to free space can be established when a room is well designed and operated at high enough frequencies for the far field conditions to be valid. In practice, even in the best anechoic chamber, only a relatively small part of the room will be free from reflections from the walls which means that the EUT and sensing antenna must be carefully positioned.

Radiated immunity testing can be performed within an anechoic chamber; an antenna being used to produce the incident field. The chamber will provide the necessary shielding to prevent the test being a nuisance to other equipment or broadcast users. As the environment approximates to free space, the fields generated can be calculated using the known free space antenna factors, provided the EUT is in the far field. Under these conditions a uniform field will be generated.

Semi-anechoic chambers are similar to fully anechoic chambers except that the floor is not lined with absorber. Such chambers suffer from reflections from the floor but are less resonant than an empty screened room. If the floor is metallic then the semi-anechoic chamber has a similar performance to an OFTS and the measurement antenna must be scanned over a range of heights. This means that the chamber must be of sufficient height to carry out the required scan, 5.5 m in the case of 10 m EUT-antenna separation. Semi-anechoic chambers have been built which give a 'site attenuation' which is within the limits specified by the FCC [Hemming[17]], but these are large and have absorber which is 1.8 m deep and hence very expensive. IBM have a semi-anechoic chamber built beneath their OFTS at Hursley Park, although this chamber does not meet all the ideal requirements, it has been empirically demonstrated over a period of months that measurements made on the OFTS correlate closely to those made in the chamber.

Semi-anechoic chambers are also used for radiated immunity measurements as the lack of reflections from the walls means that fields generated by an antenna can be considered to be plane waves and the fields reasonably uniform, although they still do vary depending on the size of the EUT. The same problems exist at frequencies below anechoic performance as in a screened room. Large semi-anechoic chambers are used by the automotive industry, examples being the Motor Industry Research Association (MIRA) facility and the Rover Group facility at Gaydon. Fields up to 200 V/m are generated within these facilities.

8.5 Conclusions

The technical limitations of the OFTS, the screened room, anechoic and semi/anechoic chambers have been discussed. Practical examples of the OFTS have also been considered in detail. A number of conclusions may be drawn:

1) to reduce the potential variation in OFTS performance, standard constructional guidelines should be formulated.

2) the current assessment of site performance, 'site attenuation', is of limited value in assessing the 'quality' of a site. Because it relies on two antennas being placed at the elliptical site focii, between which will be inherently good coupling, the site itself has little effect on the measurement. Additional measurements, based on for example the CNE, should be introduced to identify site imperfections.

3) the screened room, whilst potentially affecting radiated emission and immunity measurements because of reflection and resonance behaviour, offers a cost effective test facility for pre-OFTS radiated emission measurements, conducted emission measurements, ESD testing, conducted immunity tests and radiated immunity tests (provided that the uniformity of the field can be maintained).

4) anechoic performance and the simulation of free space conditions can be achieved if a sufficient budget is available.
5) the semi-anechoic chamber can offer a similar performance to an OFTS.
6) for immunity testing of large equipments *eg*. motor vehicles, only the semi-anechoic chamber can provide a suitable environment when taking account of budgetary considerations.

References

1. W S Atkins 'The UK Market for EMC Testing and Consultancy Services', 1989
2. D.M.Imeson 'The EMC Manager View of 1992', EURO EMC '90 Conference Proceedings, London, October 1990
3. CISPR Publication No. 22 'Limits and methods of measurement of radio interference characteristics of information technology equipment', IEC, 1985
4. FCC, 'Part 15 - Radio Frequency Devices', Code of Federal Regulations No. 47 Vol. 1, Office of the Federal Register National Archives and Records Administration, 1989
5. VDE 0871/6.78 (DIN 57 871) 'VDE Specification Radio Frequency Interference Suppression of Radio Frequency Equipment for Industrial, Scientific and Medical (ISM) and Similar Purposes', English Translation by EMACO EMC Consultants
6. EN 55 022: 1987 (BS 6527:1988) 'Limits and methods of measurement of radio interference characteristics of information technology equipment', British Standards Institute, 1988
7. Bulletin OST55, Federal Communications Commission, Washington, August 1982
8. VDE 0877 part 2 'Measurement of radio interference Measurement of radio interference field strength', Feb.1985
9. E.L.Bronaugh 'Comparison of 4 open area test sites', 7th International Symposium on EMC, Zurich, March 1987, pp 339-345
10. BS EN 55 011: 1991 'Limits and methods of measurement of radio disturbance characteristics of industrial, scientific and medical (ISM) radio-frequency equipment', British Standards Institution, 1991
11. W.Scott Bennett 'Error Control in Radiated Emission Measurements', EMC Technology 1982 Anthology, pp 171-177
12. M.Robinson and A.R.Nash 'Construction of an Open Area Test Site' EURO EMC '90 Conference Proceedings, London, October 1990.
13. R.F Harrington 'Time harmonic electromagnetic fields', McGraw Hill, 1961 pp 66-81
14. B Keiser 'Principles of EMC', Artech, 3rd edition, 1987, pp 353-355
15. L Dawson and A C Marvin 'Alternative Methods of Damping Resonances in a Screened Room in the Frequency range 30 to 200 MHz', IERE Sixth International Conference on Electromagnetic Compatibility, University of York, September 1988
16. E.E.Donaldson, W.R.Free, D.W.Robertson and J.A.Woody 'Field measurements made in an enclosure', Proc IEEE, vol 66, no.4, April 1972, pp 464-472
17. L.H.Hemming 'The chamber factor: A method of correlating to the open field site', IEEE 1983 International Symposium on EMC, pp 408-412

9

Implications of Directive 89/336/EEC

This chapter should be read by those considering investment in EMC test facilities, those requiring a view of the EMC testing and consultancy market place and those concerned with achieving compliance for 'one-off' or large equipments. The UK EMC testing and consultancy infrastructure is considered with reference to the 1989 W S Atkins report for the DTI, the 1993 Moore survey and current (March 1995) market indicators. A case study is presented for a large RF powered woodglueing machine and implications for manufacturers of large or 'one-off' products derived. The reader will acquire an appreciation of the potential effects of the Directive on the EMC market, the demands placed upon the UK EMC test/consultancy infrastructure and a critical awareness of the factors affecting product compliance with the EMC Directive.

9.1 Introduction

The EC Directive on EMC [89/336/EEC[1]] has major implications for the whole of the UK electrical and electronics industry as it applies to the vast majority of electrical and electronic products placed on the market. The definitions of components (sub-assemblies), apparatus (equipment), systems, installations and the scope of the equipment covered have been clarified by the European Commission's explanatory document[2] (Chapter 3). *Note*, this equipment must comply with the 'essential protection requirements' defined in Article 4 of the EMC Directive[1] and as described in 2.2.

To demonstrate compliance with these requirements, equipment may be tested to relevant standards (self-certification), or a technical construction file prepared for submission to a competent body. Having self-certified or obtained a certificate or report from a competent body, a manufacturer (authorised representative or importer, Chapter 3) can prepare a certificate of conformity, place the CE marking on his product and market it freely throughout the European Economic Area.

The principal assumptions for achieving compliance are that appropriate standards are in place for all types of equipment and that the adjudication of technical construction files will be carried out in a uniform and competent manner throughout Europe. As already described in Chapter 5, product specific standards covering all products are a long way off, the generic standards for the 'residential, commercial and light industry' environment are in place, but apart from prETS 300 127[3] and some provision in EN 55 011[4] and EN 55 022[5] there are no standards in place for large equipments, although the generic standards for the 'industrial' environment may be applicable to some equipments but *in situ* testing is not permitted [EN 50 081-2[6] and prEN 50 082-2[7]]. The qualification to be a competent body in the UK includes NAMAS accreditation as indicated in 2.4.2 and discussed in Chapter 3. The main concern here is that NAMAS accreditation is essentially a quality assurance scheme, similar to BS 5750[8] (ISO 9000), and does not necessarily imply that an organisation possesses the appropriate technical expertise. The DTI circulated a document in 1991, prepared by a NAMAS working group, outlining the requirements of the *technical construction file* (TCF), in a revised form this is published as a 'Guidance Document on the preparation of a TCF as required by EC Directive 89/336[20]'. Whereas previously the general understanding was that the competent body would prepare the TCF, the emphasis in this document is upon the *manufacturer* preparing his TCF (albeit in consultation with a competent body) and then submitting it to the competent body for approval. This is considered to be more 'in the spirit' of the idea of self-certification. The European Organisation for Certification and Testing, EOTC, has been established and is in the process of formulating rules for testing organisations throughout Europe [EC Green Paper on the development of European standardisation, 1990[10]]. The EOTC is concerned with all forms of testing not just EMC and is organised on a committee basis like the IEC. One of its committees is EMCIT, the committee responsible for EMC testing of IT equipment. The formation of the EOTC should produce a more uniform approach to quality assurance and technical competence of testing bodies which may become competent or notified bodies under the various Directives.

In addition to defining the requirements to be placed on manufacturers, the EMC Directive also defines the responsibilities of the 'administration' in a member state. In the UK enforcement for other than radiocommunication equipment will be the responsibility of the weights and measures authorities and the Department of Economic Development in Northern Ireland [EMC

Regulations[9]], *complaint driven* organisations. This does not seem to be in the spirit of the EMC Directive! It is not clear how other member states intend to enforce the EMC Directive, however it is reasonable to speculate that Germany will maintain its existing VDE approvals testing facilities but that they will be redirected to random testing of both indigenous and imported products. Indeed provisions have been made in the German legislation to raise a levy from transmitter operators in Germany to cover the cost of implementing the act [BSI THE, 1995[11]]. Thus whilst it may be possible to market non-compliant equipment in the UK until a complaint is made, the same product could be the subject of random testing in another member state and be found to be non-compliant. Hence for an unscrupulous manufacturer or importer, choice of member states for distribution of a product may enable him to trade without complying with the EMC Directive if he is prepared to run the risk of complaints being made or a competitor bringing his product to the attention of the enforcement agency!

The standards situation will be illustrated by a case study and the demands placed by the EMC Directive on the UK EMC testing infrastructure will be discussed with reference to the W S Atkins Report[12], the survey published by John Moore Associates Ltd[13] in June 1993 and market indicators from the results of questionaires circulated to the EMC Clubs by the York Electronics Centre and by the Institution of Electrical Engineers.

9.2 A Case Study - RF Equipment for the Woodworking Industry

9.2.1 Introduction

This case study illustrates some of the problems associated with testing a physically large item of electrical/electronic apparatus in order to achieve compliance with the objectives of the EMC Directive.

Calder Woodworking was a small company traditionally trading in reconditioned secondhand woodworking machinery. This included RF powered wood glue-curing machines. This division of Calder had a turnover in excess of £0.25M in 1988, which included the sales of new RF equipment developed and built by the firm. The company was unusual for its size as it was aware of the EMC Directive and intended to ensure its products would comply with the requirements well before the EMC Directive was due to be enforced.

The York Electronics Centre carried out emission measurements on an example of a Calder machine in June/July 1989 and the findings were published at EUROSTAT '89 [Marshman and Dawson[14]]. Since carrying out the tests Calder has become a part of Wadkin and the RF woodglueing machines are now sold under the logo of Wadkin RF.

9.2.2 RF woodglueing machinery

RF woodglueing machines consist of an RF generator with power rating of between 1.5 and 12.0 kW, supplying power to the plattens into which the pre-glued timber is loaded *via* a feeder and discharged *via* a second similar mechanism, Figure 9.1. Hence the machines are physically large, typically 7.3 m (24 ft) long by 1.5 m (5 ft) wide for the wood handling portion. The RF generator is some 2 m high by 1 m by 1 m and is a free running single valve oscillator. The whole machine weighs approximately 20 tonnes.

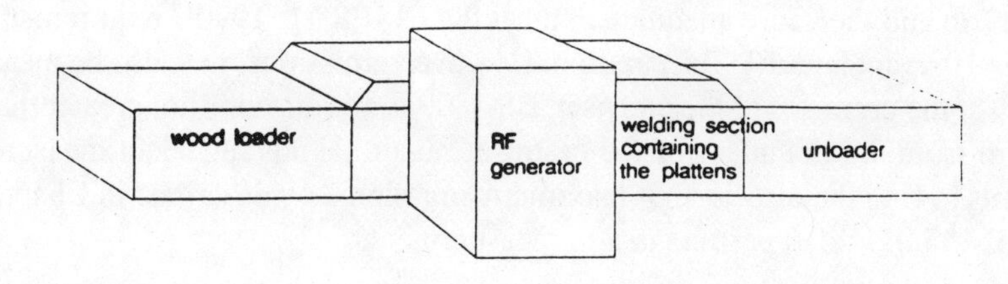

Figure 9.1 ***Arrangement of RF woodglueing machine***

Essentially the equipment consists of a relatively small RF generator connected to a large potential antenna formed by the plattens and loading/unloading mechanisms. The operating frequency of the RF generator is 13.56 MHz and is an ISM frequency defined by the ITU and published in CISPR 11[15] and EN 55 011[4]. When the tests were performed for Calder the European implementation of CISPR11, EN 55 011, had not been published. The British Standard covering this type of equipment was BS 4809: 1972[16] and was applicable to RF heating equipment, this has now been withdrawn and superseded by BS EN 55 011[4], published in 1991.

9.2.3 Testing for radiated emissions

The method for measuring the radiated emissions was given in BS 4809[16]. For frequencies between 30 MHz and 1 GHz this specified a test distance of 30 m between the centre of the source and the sensing antenna and called for the equipment under test (EUT) to be rotated to find the direction of maximum radiation from the EUT. If the equipment was too large to be rotated the sensing antenna was to be moved around the source to find the position of maximum radiation. BS 4809[16] stated that the test site should be a level area free from appreciable wave reflecting surfaces.

EN 55 011[4] the CENELEC implementation of the extensively revised, CISPR 11: 1990[15], calls for an elliptical open field test site (OFTS) as defined in EN 55 022[5] (Figure 6.3) and described in Chapter 8. Also the draft specification for measuring emissions from large telecommunications equipment, prETS 300 127[3], has become available since the Calder tests and defines a site diameter of 30 m for tests where the sensing antenna is moved around the EUT.

According to Moore[13] there is only one site capable of accommodating equipment of this size and meeting the CISPR 11: 1990[15] requirements, whilst a further 5 sites have a length in excess of 30 m but only have breadths between 7 and 12 m. Significantly the Moore[13] survey reveals that of the 29 test houses interviewed only 6 appear to have sites with a breadth in excess of 17 m and therefore adequate to meet the CISPR 11: 1990[15] requirements for a 10 m antenna/EUT separation. However, emissions may also be measured at the premises of the end user, BS 4809[16], at a distance not greater than 30 m from the boundary of the premises, again taking sufficient measurements to find the direction of maximum emissions. As described in Chapter 6, EN 55 011[4] also permits *in situ* measurements.

9.2.3.1 The site used for the measurements

In order to make a preliminary assessment of the radiated emissions from the RF glueing equipment, it was decided to test a machine at Calder's premises. The only site available for testing the equipment was the car park/loading bay behind the factory. A plan of the site (Figure 9.2) shows that the area of open space was approximately 45 m x 30 m and was surrounded on two sides by the factory buildings. The third boundary was supplied by a hill some 30 m high rising at an angle of approximately 45° and the last boundary was an area of rough ground between the hill and the factory buildings.

The ambient electromagnetic environment was measured and found to be low enough to carry out the measurements. A typical measurement is shown in Figure 9.3. A petrol generator was used to provide a separate power supply for the test equipment to minimise the effect of any conducted emissions on the measuring equipment.

9.2.3.2 Radiated emission measurements

It was not possible to carry out a full 360° coverage so measurements were made in the positions shown in Figure 9.4. It was assumed that the radiation pattern would be symmetrical about a line though the centre of the generator and plattens. The actual positions used for the measurements were dictated in part by the test site. These measurements indicated that the equipment does act as a directional source, but that the direction of maximum emissions depends on the emission frequency.

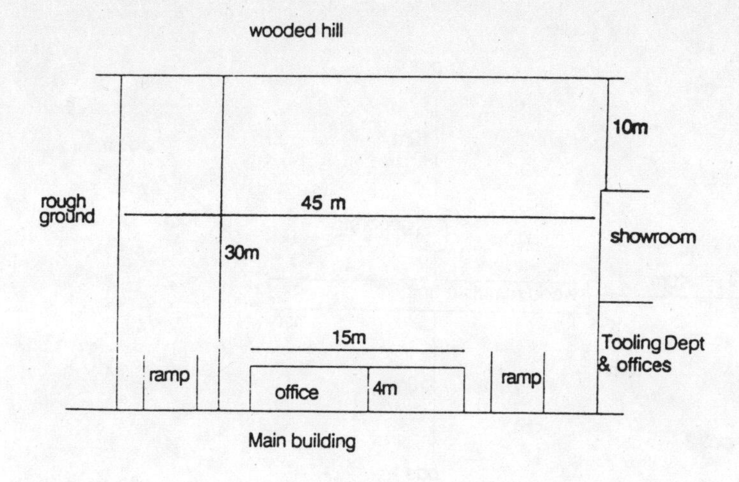

Figure 9.2 ***Plan of the site used for the tests***

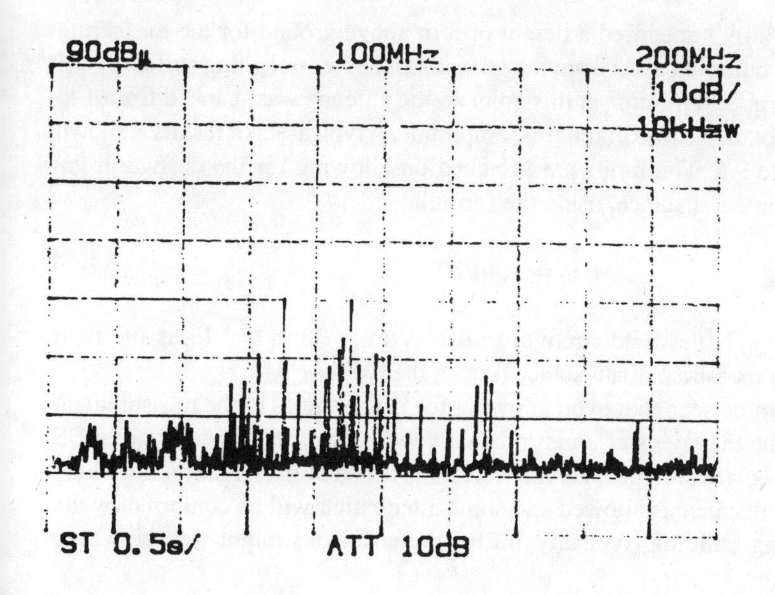

Figure 9.3 ***Typical ambient measurement***

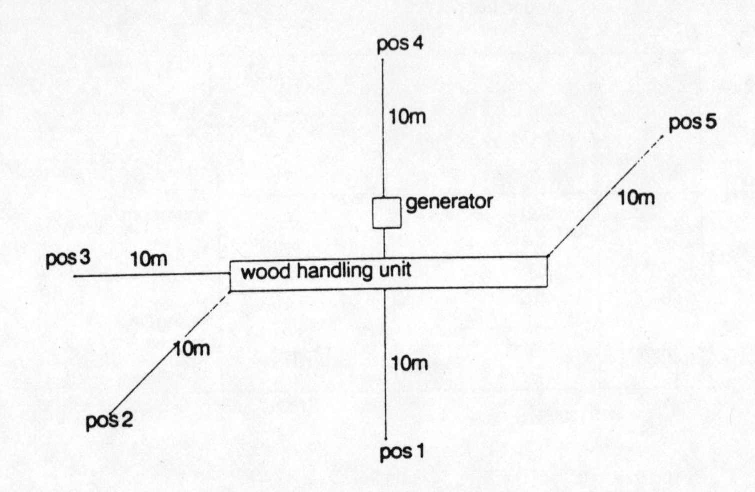

Figure 9.4 Positions from which measurements were carried out

BS 4809[16] specified a height of 3 m above ground for the measuring antenna but reflections from the ground (and other reflecting surfaces) can produce a field minima at this point so the antenna was scanned from 1 to 4 m to obtain the maximum field amplitude. A typical set of results is shown in Figure 9.5. The limits are sketched on allowing for the change in the measurement distance, using the formula:

$$E_{30} = E_d(d/30)$$

where E_{30} is the field strength in dBmV/m at 30 m and E_d is the field strength measured at a distance d from the EUT

It can be seen that even allowing for inaccuracies in the measurement technique the levels of emissions were well above those permitted in BS 4809[16], however if the machine were to be tested at an end user's site where a larger distance is allowed and some attenuation will be conferred by the enclosing building (typically 10dB), the emissions might well be within the limits.

These measurements are certainly suitable for providing preliminary test data allowing design improvements to be investigated and with some refinement may be suitable for inclusion in a TCF if this route to claiming compliance with the EMC Directive has been selected.

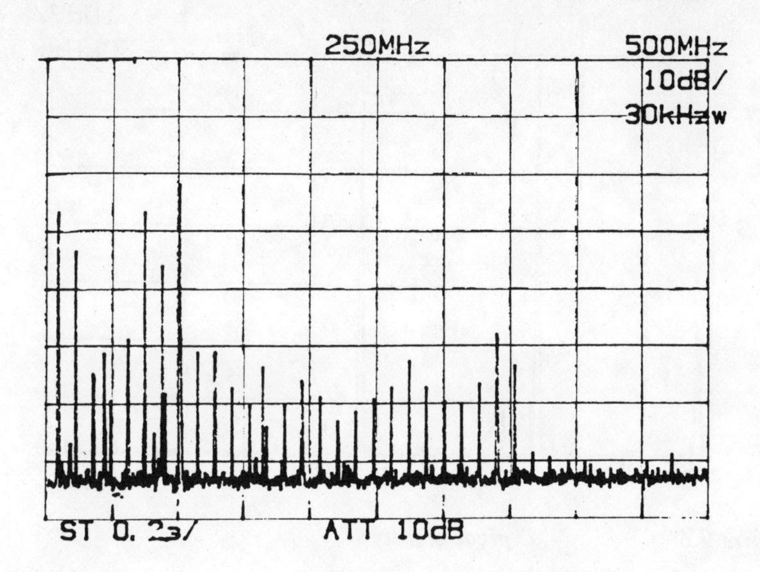

Figure 9.5 ***Typical results***

9.2.4 Conducted emission measurements

BS 4809[16] specified limits for the terminal voltage at frequencies between 0.15 and 30 MHz. The method of measurement was defined by BS 4809[16] using a voltage probe, [this is included in EN 55 011[4] and shown in Figure 6.8]. The measurements were not however carried out in this manner, but were made using a LISN, as specified by VDE 0871[17] for ISM equipment which provides a measurement of the RF currents emitted into a defined load ensuring repeatibility. Typical results are shown in Figure 9.6.

The emissions were approximately 70 dB above the Class A limit specified in VDE 0871[17], which under existing regulations allows an individual permit for the equipment tested.

9.2.5 Review of results

The measurements carried out showed that the emissions from the equipment for both conducted and radiated emissions were greatly in excess of the specified limits. Calder RF systems have reviewed their designs particularly with regard to cable layout, shielding and mains filtering. For current equipment RF generators and the plattens are enclosed in a steel enclosure with well fitting interlocked doors.

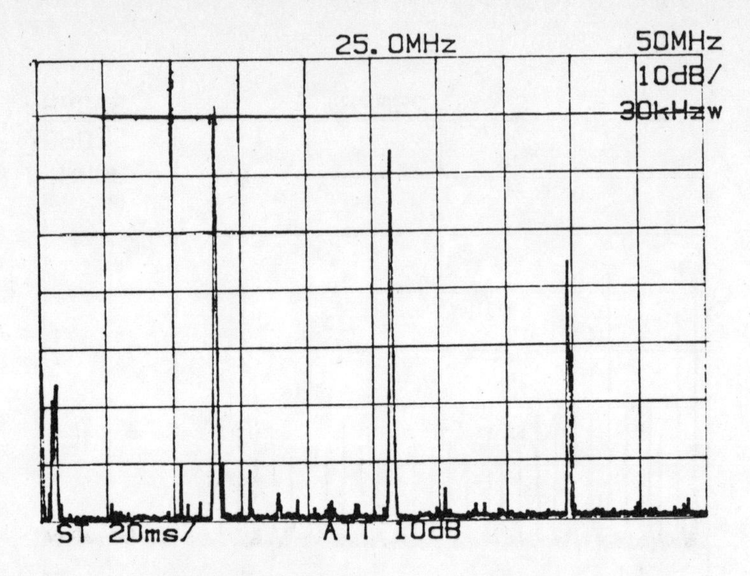

Figure 9.6 ***Typical results***

9.2.6 Immunity considerations

At the time of the Calder tests the immunity of the woodglueing equipment was not measured and there were no available immunity standards. However consideration was given to what would be regarded as a failure of this equipment. Most electronic apparatus when subjected to susceptibility testing is deemed to have failed when its performance is degraded such that its operation is outside its functional specification. For example in the case of weighing machines when the displayed weight is outside a permitted tolerance. For this type of machine a functional failure would be deemed as the machine's inability to fully cure the glue between the wood laminations. However its EMC performance may well have been substantially degraded prior to any apparent effect on its primary function.

As previously mentioned the RF generator is a single valve oscillator designed to run at 13.56 MHz. Within the ISM frequency band 13.553 to 13.567 MHz, BS 4809[16], the machine is allowed to emit high levels of radiation, however outside this band the levels of permitted radiation are restricted. Hence should the bias voltage of the oscillator be affected by either impulsive mains power supply disturbances or from the effects of external radiated fields being picked up by the power cables (very efficient antennas at these frequencies) the oscillator may run 'off tune' and the operating frequency may be outside the guard band. Under these conditions a very efficient RF power source operating at a non-ISM frequency exists and thus provides a potential threat to broadcast or other services.

It is apparent from the above discussions that susceptibility testing for this and other ISM equipment should be considered carefully and in some instances should be carried out concurrently with emission measurements.

The only applicable immunity standard for this type of equipment is the generic standard for the 'industrial' environment, this exists in draft form, prEN 50 082-2[7] and appears to provide little guidance as to how high power equipment and large installations should be tested for immunity. Available mains disturbance equipment is also limited typically to EUTs with power ratings of less than 10 kW. Surge generators are available for use with equipments of up to approximately 20 kW rating.

Radiated immunity measurement techniques for large equipments have been developed by the automotive and aircraft industries. In the case of the automotive industry this involves illuminating a vehicle with EM waves within a large semi-anechoic chamber, Chapter 8. Such measurements could be performed on a woodglueing machine but would be prohibitively expensive to a company such as Calder. Costs are of the order £2 k per 8 hour shift in addition there would be transportation and setting up costs. Alternative techniques are required such as partial illumination of the likely susceptible parts of the equipment, provision for this is included in the proposed revision to IEC 801-3 [90/29283 DC[18]].

Alternatively a manufacturer may choose not to perform this test when using the generic standard as long as he can justify his actions. In this case he may be able to claim that the shielding employed to limit the emissions will act in a reciprocal manner and therefore radiated immunity testing is not required.

9.2.7 Compliance — design considerations

Conducted electromagnetic energy can be limited by using appropriate filtering techniques on power and control cables. Radiated emissions from RF heating installations is the most likely reason for non-compliance with the EMC Directive [Rowley, 1991[19]]. EN 55 011[4] specifies limits at a distance of 30 m from an external wall of a building housing the installation (*see* Chapter 6). If the emissions exceed these limits then the equipment does not meet the standard and does not comply with the EMC Directive.

All radiated emissions may be limited by enclosing the equipment in an efficient RF screen. Alternatively the equipment design can be improved by ensuring that all the RF energy is generated at the fundamental frequency and that this frequency remains within its ISM band. Shielding is then only required to meet health and safety requirements. Since shielding effectiveness is reduced at higher frequencies it is important to limit the production of higher harmonics. This may require the addition of filters on the output from the generator [Rowley, 1991[19]].

The RF woodglueing machine is an example of a conventional open loop RF heating system, its operating frequency is dependent on the generator and its interaction with the load to be heated or dried. For example if the timber has a higher moisture content than normal, then the dielectric constant will also be changed and the capacitance of the load formed by the plattens and the timber/glue will be different and hence the tuned frequency of the system will also change. New designs of RF dielectric heating systems use crystal controlled oscillators and have a fixed output impedance, which makes the system independent of load. A closed loop control system can also be used to re-tune the system constantly and maintain the operating frequency within the ISM bandwidth [Rowley, 1991[19]].

9.2.8 Conclusions

Some of the problems to be faced by manufacturers of physically large electrical/electronic systems when trying to comply with the requirements of the EMC Directive have been illustrated by this case study. These may be summarised as follows:

Radiated emission measurements

i) • it may be prohibitively expensive to test a large system at a sufficiently large test facility especially for a small company such as Calder RF Systems (typical of a 'small to medium size enterprise' or SME).

• the Moore[21] survey indicates that the choice of a suitable test site may be may be severely limited in any case.

• the method described for determining the radiated emissions, with some refinement, is believed to offer a reasonable technique for radiated emission assessment and the results could be included within a Technical Construction File to demonstrate compliance. Certainly this technique provides a cost effective preliminary guide, before the expense of transporting such a system to an approved test site is incurred.

• testing at an end user's site is also permitted by EN 55 011[4].

ii) for equipment operating on an ISM frequency it may be desirable to carry out susceptibility and emissions testing concurrently as the failure of a system undergoing susceptibility testing may result in a change to the emission performance of the system.

iii) it is unlikely that the generic immunity standard for the 'industrial' environment prEN 50 082-2[6] can be applied to this equipment as *in situ* measurements are not permitted by the generic standards.

iv) commercially available mains disturbance test equipment has limited power handling capability.

v) the problem of testing the radiated immunity of a large system is likely to be prohibitively expensive for an SME and alternative techniques need to be investigated. Possible techniques applied to diesel generating sets are discussed in Chapter 10. An alternative to applying the generic immunity standard, would be to justify that radiated emissions and radiated immunity should be reciprocal for this equipment and will be a function of the screening and therefore radiated immunity testing does not need to be performed. With the generic standards it is the manufacturer's responsibility to choose which tests should be performed in order to demonstrate compliance with the protection requirements of the EMC Directive.

9.3 The EMC Testing and Consultancy Infrastructure in the UK

In June 1989 W S Atkins Management Consultants published a report on behalf of the Department of Trade and Industry (DTI) entitled 'The UK Market for EMC Testing and Consultancy Services'[12]. This report provided an estimate of the number of different products on the UK market which would be required to comply with the EMC Directive, an inventory of EMC capabilities, facilities and resources among UK test houses and an indication of the mismatch between demand for and supply of EMC services. This study was carried out during the period February and March 1989.

The author of this book carried out a limited survey in October 1990 on behalf of the York Electronics Centre (YEC). This survey was intended to provide an indication of the number of products to be tested, the demand for testing and consultancy services and the potential demand for low cost test equipment and test facilities.

The latest survey of testing facilities was carried out by John Moore Associates Ltd[13] and published in June 1993.

In addition, the YEC circulated questionaires to the EMC Clubs and the IEE to its members, the results from these gave some market indicators at the beginning of 1995.

9.3.1 Demand for EMC testing and consultancy services

The Atkins[12] report gave the following estimates:

i) Product types — 2,300 specific product types on the UK market required to comply with the EMC Directive
ii) UK manufactured products — 7,400
iii) Products imported into the UK from outside the EC — 2,700
iv) Estimate of the number of products requiring testing — all UK manufactured products plus 50% of imported products multiplied by the number of product variants (average 3) — total number of equipments to be tested — 26,000.

The Atkins[12] report detailed the assumptions made in estimating the total testing and consultancy times required for each product type, *ie* whether self-certification and therefore testing to product specific standards can be carried out, or whether a Technical Construction File (TCF) will need to be prepared. In addition estimates were made for failure rates. All times were based on 'shifts' (8 hour period at an average cost of £1000). Consultancy effort was assumed to be 25% of the time required for the appropriate test, an 8 hour consultant day at £500/day was also assumed. From the estimated numbers of products to be tested and the above assumptions, it was estimated that the demand would be 150,000 test shifts and 50,000 consultancy shifts to clear the product inventory at that time (1989). The total cost was estimated at £150 M and £25 M for testing and consultancy respectively. This is summarised in Table 9.1 reproduced from Atkins[12].

Table 9.1 — Demand for EMC Testing and Consultancy Services

Testing (facility shifts)	**Emissions**	**Immunity**	**Total**
Technical file	54,381	38,961	93,342
Self-certification	23,646	37,658	61,304
Total	78,027	76,619	154,646

Consultancy (man shifts)	**Emissions**	**Immunity**	**Total**
Technical file	13,944	8,563	22,507
Self-certification	11,363	14,212	25,575
Total	25,307	22,775	48,082

Moore[13] suggests that by 1993 the testing demand had fallen to 'somewhere between 40,000 shifts and 50,000shifts.' This is accounted for in a number of ways:

i) the delay in implementation has allowed manufacturers time to adopt good EMC practices, resulting in a reduction in the number of test failures and therefore the perceived demand for retesting.

ii) their has been a steady investment in new facilities.

iii) the increase in standards availability has been accompanied by investment in in-house test facilities as companies have pursued the self-certification route, hence reducing the demand on third party facilities.

iv) the DTI's awareness campaign has been 'helpful in extending knowledge and provoking appropriate responses....'

v) the number of NAMAS accredited test houses has increased, most have been granted Competent Body status by the DTI and hence the availability of facilities for performing TCF work is increased.

9.3.2 Supply side — availability of testing and consultancy services

Atkins[12] carried out a survey amongst 37 UK testing laboratories having EMC capabilities and calculated the potential capacity to supply EMC services in terms of shifts. This capacity was split into Open Field Test Site (OFTS) emission testing and screened room/anechoic chamber immunity testing (an allowance was made for some emission testing to be carried out within these latter facilities). From the gross capacity calculated the net capacity was estimated by taking account of the capacity allocated to defence related work.

Similarly the capacity of consultancy services was estimated. Assumptions made were that: consultants would be available for only 1 shift per day, 50% of a consultant's time is actually administration, supervising testing, marketing and other management functions.

The main points highlighted by the survey of the 37 laboratories were:

i) 17 were in-house testing facilities for their parent companies
ii) there were 95 screened enclosures and anechoic chambers in the UK but only 25% were fully anechoic
iii) there were 26 OFTSs, 60% were covered and weatherproof
iv) an average of 40% of the capacity was devoted to defence related work
v) capacity utilisation averaged 75%

Atkins[12] analysis yielded a capacity of 26,000 testing shifts to meet the demand of EMC Directive related testing and 8,000 consultant days to meet the demand for consultancy services. Only 20% of the total testing capacity was suitable for emission testing because of the shortage of OFTSs. The results are summarised in Tables 9.2 and 9.3.

Moore[13] estimated the EMC testing resources available in 1993 and these have been included within Tables 9.2 and 9.3. These show that compared with 1989 the overall growth in screened enclosures, semi-anechoic and anechoic chambers is of the order 47%, the growth in OFTS 35% and the growth in staff 39% (these do not tally exactly with Moores estimates, but indicate the trend). Moore[13] finally concludes that the number of chambers is between 160 and 170 and the number of OFTS is 40 with an equivalent capacity of 35,000 shifts (single shift working).

Table 9.2. EMC Laboratories' Facilities, Resources and Capabilities. Source W S Atkins and J Moore Associates				
Laboratories providing services	**Third parties**	**In-house**	**All**	**1993**
No. of companies sampled	20	17	37	41
No. with NAMAS accreditation	8	1	9	38
No. undergoing NAMAS	3	4	7	5
Facilities				
No. of screened enclosures	42	20	62	90
No. of semi-anechoic chambers	4	4	8	36
No. of anechoic chambers	15	10	25	14
No. of OFTS	15	11	26	35
No. of test personnel	134	84	218	}
No. of consultants	72.5	42	114.5	}447

Table 9.3 Capacity to supply EMC Testing and Consultancy services (shifts)

Source: W S Atkins/J Moore Associates

	OFTS (emissions)	Screened enclosures/ anechoic chambers (emissions/immunity)	Total	Moore estimate 1993	Consultant Total
Independent labs	3143	12414	15557		5693
In house labs	2809	7861	10670		2703
Total	5952	20275	26227	35000	8396

9.3.3 Comparison of demand for and supply of EMC testing and consultancy services

Table 9.4 compares demand versus supply for EMC testing and consultancy services, again reproduced from Atkins[12] but with the addition of the Moore[13] 1993 estimates.

Table 9.4 Demand **versus** ***Supply for EMC testing and consultancy services***

	Testing (facility shifts)				*Consultancy (man shifts)*
	OFTS (emissions)	*Screened enclosures / anechoic chambers (emissions/immunity)*	*Total*	*1993 Total*	*Total*
Demand	73,131	81,515	154,646	50,000*	48,082
Supply	5,952	20,275	26,227	35,000	8,396
Supply as % of Demand	8 %	25 %	17 %	70 %	17 %

Source: W S Atkins and J Moore Associates

* assumes no backlog

In 1989 Atkins concluded:

i) demand exceeded supply by a factor of 12 for open field emission testing

ii) demand exceeded supply by a factor of 4 for immunity and emission measurements performed within a screened enclosure or anechoic chamber

iii) demand exceeded supply by a factor of 6 in testing taking emissions and immunity together

iv) demand exceeded supply by a factor of 6 in consultancy. Atkins[8] only claimed a +/- 50% accuracy for these results because of the large numbers of assump tions made and suggested that the mismatch lay in the range 3 to 9.

Figure 9.7 shows the Atkins market projections for EMC testing services. This shows that immediately following implementation there will be a large number of products which require to be tested in order to demonstrate compliance. Subsequently the demand for services will only be required for new products entering the market. The projections indicate that the accumulated backlog of products requiring testing will not be cleared for 15 years, assuming a 10% annual growth in services! The potential implication here is that a manufacturer may have to choose between delaying a product launch or launching a product for which compliance has not actually been demonstrated.

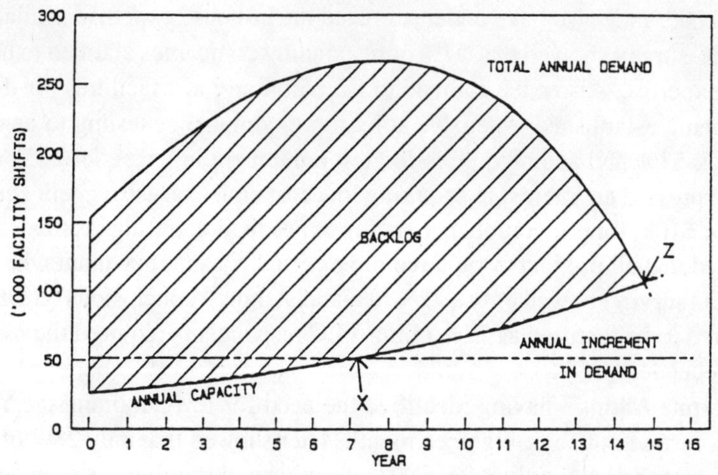

Figure 9.7 ***Market projections for EMC testing services 1989***

The Atkins[12] report also drew the conclusions that there was an urgent need for *investment in* testing facilities, particularly *OFTSs,* where the mismatch is the most severe and in the *training of EMC engineers*. The availability of suitably trained personnel was thought to be the *'key limiting factor'* on the expansion of capability. Atkins[12] also highlighted that there was an inherent difficulty regarding investment in facilities and resources because of the scale of the 'start-up' demand compared with the longer term utilisation, when the supply/demand balance will be radically different. This latter factor was clearly a major reason for the DTI support for extending the transitional period to four years in order to provide a 'soft-start' to the compliance testing programme.

The YEC 1990 survey provided complementary information which to some degree supported Atkins[12] conclusions. This survey was probably distorted by the high proportion of responses from companies with a degree of EMC awareness. Reviewing the numbers of products to be tested (and making allowance for misunderstanding of the Directive) out of the 47 replies received, it was believed that 1974 products would require testing. One of the responses, however, was from a company manufacturing artificial limbs who assumed that every product would require testing. In reality the TCF route to compliance may allow generic tests to be performed and thus the total number of products was reduced to 1474 from the 47 companies. The Atkins[12] total of 7400 for the whole of the UK must be considered as a serious underestimate when compared to this result from the YEC survey. However Atkins[12] assumed that all products would be assessed whereas the YEC survey showed that some 45% of products had had their EM performance assessed, hence indicating a reduction in the load on test houses and consultancy.

Atkins[12] may also have underestimated the in-house expertise available: the YEC survey showed that 57% of responding companies claimed to have EMC expertise. Also 51% claimed to have in-house test facilities (or these were being established) to enable at least pre-compliance testing to be carried out. 57% had acquired emission test equipment in excess of £25k and 72% expressed an interest in acquiring low cost emission assessment equipment (< £10k) for pre-compliance testing or production testing. Atkins[12] also assumed that all products would require external consultancy input whereas the YEC survey showed only a 72% demand. Atkins[12] conclusions on OFTS appeared to be justified as 77% of the YEC respondents required the use of an independent OFTS.

Despite Atkins[12] having identified the need for EMC training the YEC survey carried out some eighteen months later showed that only 28% of the respondents had any policy for EMC education and training. As more engineers become educated in the discipline of EMC then the demand for consultancy services will drop and as new designs implement EMC practices, the duration of testing will also reduce.

The Moore[12] report demonstrates a fall in demand for services accounted for as described in 9.3.1 and indicates that the Atkins[12] backlog may have been significantly reduced although no estimates are given. In practice the demand for EMC testing services is increasing as 1995 progresses and the nearer 1 January 1996 looms! Quantitative assessment for the demand on independent test laboratories is impossible to estimate given the volatile market state. Undoubtedly the sales of low cost pre-compliance EMC test equipment are increasing, but there are considerable indications that the numbers of products to be tested have been seriously underestimated. Of concern is that the demand for EMC training does not appear to be tracking the aquisition of pre-compliance test equipment with the implication that manufacturers may not have developed a comprehensive EMC strategy. The danger is that untrained personnel may not be equipped to make measurements or interpret test results correctly.

9.3.4 Market Indicators (March 1995)

The York Electronics Centre devised a questionaire for the Yorkshire and Humberside EMC Club which was established as part of the national club network under the DTIs awareness programme. This questionaire was subsequently circulated amongst the other EMC clubs and to IEE members. 146 returns were received.

The main results from this survey are summarised as follows:

1) Understanding of the EMC Directive amongst companies is still un certain, 42% claimed to have an adequate knowledge. Clarification of the EMC Directive and access to up to date information were both important issues.
2) 55% of companies claimed to have an adequate understanding of how they were affected.
3) 29% possessed an EMC strategy and plan, but developing a strategy was not high on their agenda.
4) 23% were implementing a plan or acting on their understanding.
5) 11% claimed to have compliance for their products.
6) 38% had established an EMC approach to design.
7) 19% had a plan for EMC education and training.
8) Performing and obtaining information on in-house testing was a high priority.

A number of observations were made on the results of the survey. There were marked differences in response between the EMC Club and the IEE results. In all cases the IEE responses were more negative. If it is assumed that the IEE responses are representative of companies that have not joined EMC Clubs, then it may be concluded that the clubs are being successful in clarifying and disseminating information. There were also marked differences in club responses which appeared to be regional in nature. There were also differences which might reasonably be attributed to survey timing, the questionaires were distributed over a period of seven months and it is quite clear that individuals within companies have become more focussed on entering the 12 month countdown period to 1 January 1996 when compliance becomes mandatory.

9.3.5 Conclusions

It may be concluded that there is still a mismatch between the available test facilities and the demand for product testing despite the expansion of test facilities. The effect of the extended transitional period will have reduced the backlog of testing predicted by Atkins. The backlog and the steady state testing demand are also being reduced by the establishment of in-house test facilities, whilst these may initially be established for pre-compliance work the likelihood is that as confidence grows in-house results will be used as the basis for declaring conformance.

Test houses experienced a dramatic increase in the demand for testing towards the latter half of 1994. Many companies may find themselves being forced to queue for services from mid 1995 onwards. The message is clear that action must be taken by manufacturers to plan their route to compliance and reserve test time in advance to avoid disappointment!

From the latest YEC survey it is apparent that whilst many manufacturers are equipping themselves with pre-compliance test equipment less than 20% have recognised the need to educate and train their engineers or other staff. EMC measurements are not equivalent to making measurements with a digital voltmeter or an oscilloscope and it is comparatively easy to obtain meaningless results! Education is the best investment that a company can make to control the issues arising from the need to make products compliant. If engineers have an in-depth understanding of EMC a company has a valuable resource available that will ensure that EMC is built in at all stages of company policy and product development. Investment in education will very quickly pay for itself. (For further consideration of EMC education see appendix E.)

References

1. 89/336/EEC Council Directive 'on the approximation of laws of Member States relating to electromagnetic compatibility', Official Journal of the European Communities No.139 25 May 1989, pp19-26
2. EC explanatory document on Council Directive 89/336/EEC, 111/4060/91/EN-Rev. 1, 1991
3. prETS 300 127 'Radiated emission testing on physically large systems', European Telecommunications Standards Institute, 1990
4. BS EN 55011: 1991 'Limits and methods of measurement of radio disturbance characteristics of industrial, scientific and medical (ISM)

radio-frequency equipment', British Standards Institute, 1991
5. EN 55 022: 1987 (BS 6527: 1988) 'Limits and methods of measurement of radio interference characteristics of information technology equipment', CENELEC (BSI), 1987 (1988)
6. 91/21829 DC prEN 50 081-2 'Electromagnetic compatibility - generic emission standard generic class: industrial', BSI CENELEC, 1991
7. 91/21828 DC prEN 50 082-2 'Electromagnetic compatibility - generic immunity standard generic standard class: industrial', BSI CENELEC, 1991
8. BS 5750: parts 1 to 4 '- quality management and quality system elements', British Standards Institution, 1987-1990
9. DTI 'Statutory Instruments 1992 No. 2372 Electromagnetic Compatibility Regulations', HMSO, October 1992
10. Department of Trade and Industry 'The European Commission's Green Paper on the Development of European standardisation. A United Kingdom Consultative Document', October 1990
11. Technical Help to Exporters 'Electromagnetic Compatibility Europe', BSI, 1994 ISBN 0 580 20917 2 update 6
12. W S Atkins 'The UK Market for EMC Testing and Consultancy Services', 1989
13. John Moore Associates Limited 'Electromagnetic compatibility testing in the UK', ISBN 0 9521797 0 9, June 1993
14. Dawson & Marshman, 'RF Woodglueing equipment', EUROSTAT '89 conference, London, October 1989
15. CISPR publication No. 11 'Limits and methods of measurement of electromagnetic disturbance characteristics of industrial, scientific and medical (ISM) radio-frequency equipment', IEC, second edition 1990
16. BS 4809: 1972 'Specification for radio interference limits and measurements for radio frequency heating equipment', BSI, 1972 (amended 1981)
17. VDE 0871-1 : 1985 'Radio interference suppression of radio frequency equipment, for industrial, scientific and medical (ISM) and similar purposes', VDE, 1985
18. 90/29283 DC 'Draft - revision of IEC publication 801-3: electromagnetic compatibility for industrial-process measurement and control equipment Part 3: immunity to radiated radiofrequency electromagnetic fields', BSI, 1990
19. A T Rowley 'EC Directive on EMC: Implications for radio technology', Capenhurst, October 1991
20. DTI 'Guidance Document on the preparation of a Technical Construction File as required by EC Directive 89/336', October 1992

10
Achieving Compliance with the EMC Directive

This chapter is intended for those involved in achieving compliance for a product. The design engineer, the approvals engineer, the quality manager, the EMC test engineer and the 'signatory' to the declaration of conformity appointed by a manufacturer or his authorised representative. An action plan for achieving compliance is presented as a flowchart. The use of this plan along with an up to date list of harmonised standards is illustrated by example. Examples used are: a personal computer, IEEE communications card, diesel generating sets and a distributed data gathering system for installation in a factory. The reader should be in a position to identify the route to compliance with the EMC Directive for his product, the appropriate harmonised standards and an understanding of the potential problems and pitfalls involved in achieving this compliance.

10.1 Introduction

The EMC Directive[1] and its UK implementation have been considered in some detail (Chapters 2, 3 and 4). The implications for the UK industry and UK agencies or franchises importing products from outside the European Economic Area have been discussed (Chapter 9). The current status of harmonised standards and their inadequacies have also been described (Chapters 5, 6 and 7). However, despite any uncertainties which remain the EMC Directive is legally binding in all countries of the European Union and the European Economic Area and all products, without exception, must comply by January 1996 [amending Directive, 1992[2]].

Within an organisation the approvals engineer or other designated person is responsible for achieving compliance with the EMC Directive for a particular product, equipment or system. He must therefore attempt to follow a reasonable course of action accepting the prevailing limitations and uncertainties. He must keep in mind the essential protection requirements defined by the EMC Directive which have the laudable aim of establishing an environment within which all electrical equipment can operate without interference and without interfering with the intended function of other equipment or broadcast and other legitimate spectrum users. This implies that a reasonable or pragmatic approach should be taken and the approvals engineer requires guidance in the form of an action plan.

The author's experience of providing consultancy to a wide range of industrial clients has enabled such an action plan to be developed. In this chapter the plan is applied to four examples which illustrate the difficulties that can be encountered. The examples are, a personal computer (PC), an IEEE 488 plug-in communications card for use in a PC, a diesel generating set (genset) and a distributed data gathering network used in a factory.

10.2 Action Plan for Achieving Compliance

In order to advise industrial clients over the last few years, it has been necessary to formulate a series of steps which will ultimately enable them to claim compliance with the EMC Directive. These steps are summarised as follows:

1. Define existing products which it is intended to market after January 1996. Assess the EM performance of these against applicable standards where possible and establish whether or not a technical problem for compliance with the EMC Directive is likely to exist. A commercial decision may also be required if equipment is identified as requiring additional development work, because it may be more appropriate to declare a product obsolete before the end of the transitional period and direct resources to new product development.

2. For new developments establish an EMC project strategy and an EMC Management or Control Plan. Design with EMC in mind, from the concept stage, to avoid costly retrofits and evaluate the EM performance at prototype and pre-production stages.

3. Where a lack of in-house expertise exists recourse should be made to a third party consultancy and/or test facilities and a training programme for engineering staff introduced.

4. Ensure that perceived problems can be considered by the standards making machinery, either by representation through Trade Associations or directly through BSI. If applicable standards are not available then it may be appropriate to devise suitable guidelines (in conjunction with or

through a Trade Association) which may be submitted to form the basis of a harmonised European standard. The extended transitional period has made this sort of response more feasible.

This approach allows the design engineer to enjoy the confidence of knowing that his existing equipment (or new development) has an adequate immunity and is unlikely to have an emission problem. However the approvals engineer still needs to be able to assure the 'signatory' that a declaration of conformity can be made. This requires equipment to be tested to relevant standards in order to self-certify, or requires a Technical Construction File (TCF) to be prepared including a certificate or technical report from a competent body or in the case of radio transmission equipment a type examination certificate from a notified body.

The following is a formalised plan of action for companies or organisations manufacturing or importing electrical/electronic products, equipments or systems with which compliance with the EMC Directive will be mandatory:

i) Identify equipment currently in production and which will remain in manufacture after January 1996 and *is not excluded* by the provisions of the explanatory document (Chapter 3).

ii) Identify equipment under development which will also be manufactured after January 1996 and is *not excluded* by the provisions of the explanatory document (Chapter 3). In this instance also prepare an EMC management/control plan within the project planning and ensure that the design is reviewed from an EMC viewpoint at all stages.

iii) Is there a product specific relevant standard (Euro Norm), or standards, applicable to the equipment? Refer to Tables 5.2 to 5.5, Chapter 5.

iv) Are there generic relevant standards which are applicable?

v) If the answer to these questions is YES, or likely to be YES by January 1996, then the self-certification route (defined in Chapter 2) for achieving compliance may be chosen and the equipment may be tested in accordance with the appropriate standard or standards. Testing may be carried out in-house by the manufacturer or by a third party test house. *In this instance the test house does not have to be accredited or be a competent body.*

vi) If the answer to iii) or iv) is NO, then the TCF route for achieving compliance must be used (defined in Chapter 2). In this case the manufacturer has no alternative but to submit a TCF to a competent body. Testing and/or a theoretical assessment of the equipment can be prepared by the manufacturer, along with a description of the equipment and the EMC provisions that have been made in the design. This TCF is then submitted for assessment by the competent body which will issue a certificate or technical report for inclusion within the technical file. As already discussed in Chapter 2, a competent body in the UK is likely to be a NAMAS accredited testing laboratory and preparation of the TCF and any testing required is likely

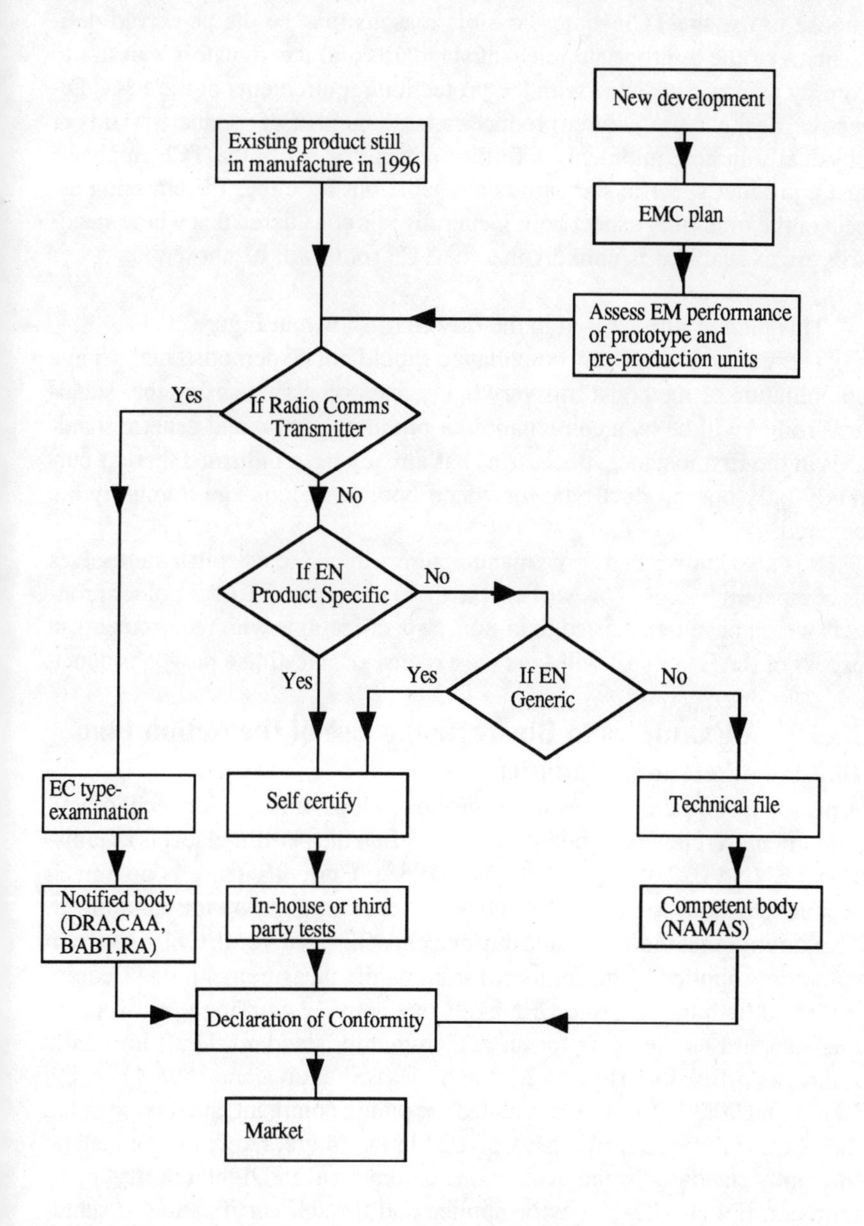

Figure 10.1 Action Plan for Achieving Compliance

to be carried out in conjunction with the chosen competent body [EMC Regulations,1992[3]]. Where standards do not exist for a particular type of equipment after the beginning of January 1996, this route will be mandatory.

vii) Should the answer to iii) or iv) be YES, a manufacturer may still choose to use the TCF route. Possible reasons may be the perceived deficiencies of the appropriate relevant standards and the signatory's desire to comply as far as possible with the protection requirements of the EMC Directive, or the manufacturer produces a large number of product variants or physically large equipment. A further reason for using the TCF might be that a product specific standard is available but for either the emission aspect or the immunity aspect only. Generally it is considered that where standards are available it is unlikely that the TCF route will be chosen.

This plan is summarised in the flowchart shown in Figure 10.1.

There is no reason why compliance should not be demonstrated using a combination of methods. It is very likely that compliance using the 'standards route' will be by a combination of product specific and generic standards in the first instance, because as has already been indicated there is currently only one product type for which both emissions and immunity are covered.

It is also known that some manufacturers intend to establish themselves as competent bodies. This will enable them to prepare TCFs for older products which have been tested to in-house specifications with requirements in excess of the ENs. This will save the expense of retesting a proven product.

10.3 Examples to Illustrate the Use of the Action Plan

10.3.1 Personal computer

A personal computer comes under the heading of 'information technology' equipment. A relevant standard exists to define the permitted levels of emissions, EN 55 022: 1987 or BS 6527: 1988[4]. Currently there is no agreed product standard to define the immunity level required by this equipment. This implies that the TCF route must be chosen. However IEC 801 has been commonly applied by the industry for immunity measurements on IT equipment and both the IEC and CENELEC are actively working towards using this standard as the basis for an ITE immunity standard. Draft immunity standards prEN 55 101 parts 2, -3 and -4[5] (BSI equivalents: 89/34172, 89/34171 and 90/30270) were circulated for public comment and it is expected that these will be adopted as EN 55 024 in the future. However the generic immunity standard for the 'residential, commercial and light industry' environment, EN 50 082-1[6] may be applied and the self-certification or standards route for achieving compliance may be adopted for this product. In addition to EN 55 022 for emissions, testing to EN 60 555-2 for harmonics will also be required as a PC may be defined as 'domestic' equipment using electronic devices.

Although self-certification may be seen as the easier and less expensive route to compliance, there are a number of potential difficulties in applying the relevant standards:

1. The system to be tested must be defined. A PC consists in its minimum configuration of the computer itself (which may include hard and floppy disk drives), a keyboard and a VDU. In addition it is likely to use a 'mouse' and it may be connected to one or more peripheral devices such as a printer or to a network. Since the interconnecting cables are very often the source of emissions or have the least immunity to external interference, the system must be tested with the cables intended to be used with the system. Hence a representative system must be chosen by the manufacturer.

2. The system is required to be exercised during testing and this will involve running a test program which is likely to produce the maximum emissions from the PC and all the peripherals connected to it. It may also be deemed necessary to operate the keyboard itself. In order to do this a keyboard actuating system, remotely operated and which will not interfere with the measurements, is required. Suitable pneumatically operated actuators have been designed and built by the larger PC manufacturers.

3. As previously discussed (Chapter 6) the definition of class A and B limits as defined in EN 55 022[4] are not clear until the footnote is read. Class A is for industrial or commercial applications and class B for domestic or residential, however there is also a rider on the use of class A which appears to be contrary to the aims of the EMC Directive, namely that there may be restrictions in some countries on the use of class A equipment. In conclusion it must be deduced that for the PC manufacturer to be able to market his product freely, it will be necessary to test to class B limits. This is actually the condition which exists for manufacturers wishing to comply with the US FCC requirements and the German VDE specifications (Chapter 12).

4. The emissions from a system may be significantly affected by the layout of interconnecting cables or the position of the mains cable. This may be significant when comparing test results obtained for the same equipment on different sites, or indeed measurements repeated on one test site.

10.3.2 IEEE 488 plug-in communications card for use in a PC

A large number of plug-in cards such as IEEE 488 interface cards are now available for use with personal computers. Reference to the explanatory document[7] and the EMC Regulations[3] suggests that sub-assemblies do not need to comply with the EMC Directive[1] unless they are marketed separately from the equipment with which they are designed to operate and have an intrinsic function. This implies that a manufacturer of a personal computer would specify to a supplier of interface cards those EMC provisions deemed necessary by that manufacturer to ensure that his end product, the PC, complies with the Directive. If however the interface manufacturer wishes to place his product on the market then it will need to comply in its own right. The difficulty here is that the interface only operates when assembled into a PC and therefore the interface manufacturer has to test the interface installed in a PC, but which PC? If he wishes to market the interface as suitable for use with a range of PCs then the implication is that he will have to test it with all PCs with which it can be used. A very costly exercise. The alternative is to accept a limited market place.

Assuming the manufacturer does intend to issue a certificate of conformity for his product, then as discussed it will be necessary to test the EM performance of the IEEE card assembled into one or more PCs. Testing required will then be as described for a PC (as described in 10.3.1) along with its attendant difficulties.

In practice the IEEE interface manufacturer is likely to test his product in one PC and record this in his declaration of conformity. He may also choose to determine, for example, the relative increase in emissions when his card is added to the chosen system, this information may be communicated to the card installer to give some indication of the likely change in the PC EMC performance if the card is installed in a PC different to that used during the testing. The onus will then be placed on the end-user should he wish to use the card in an alternative PC.

This approach is consistent with the US FCC regulations described in Chapter 12. The FCC Rules require internal 'peripherals' which in effect connect a digital device or PC to another item of equipment, to comply when tested as part of a composite system.

10.3.3 Diesel electric generating sets

Diesel Electric Generating Sets (Gensets) are manufactured with power outputs ranging from typically 10 kW to 2 MW. Applications include use as standby generating sets in the case of mains failure, peak lopping locally (for example on a factory site to prevent the kVA maximum demand being pushed into the public utility's next tariff band) and as a mains supply in remote areas or where diesel fuel is comparatively low cost. Most generating sets are specifically configured to meet a customer's requirements and although constructed from a standard range of sub-assemblies, each set invariably has differences. Most Genset manufacturers only manufacture the control equipment to their own designs and buy in diesel engines and alternators (the major capital items), which are assembled onto a suitable skid. Compared with other industries volumes are low, capital cost high and in a very competitive market place, margins are low.

When applying the 'action plan' there are currently no product specific relevant EMC standards for diesel generating sets. The generic standards for the 'industrial' environment [EN 50 081-2[8] and prEN 50 082-2[9]] must be considered but they do not cater for large items of equipment, indeed *in situ* measurements are specifically excluded by EN 50 081-2[8].

When the alternative TCF route is considered the immediate question to be answered by the manufacturer and the competent body is, whether tests need to be carried out on all Gensets or whether it is appropriate to test representatives from the range? The only acceptable answer from the manufacturer's viewpoint is to test representatives from the range, otherwise costs would be prohibitive. This can also be justified by taking the 'reasonable' point of view, although each Genset may in effect be custom built they actually consist of identifiable sub-units *viz* the diesel engine, alternator and control system and for a given power rating are likely to have similar EM characteristics irrespective of engine manufacturer or even alternator manufacturer. This reasonable action may also be justified from the viewpoint that it is the 'essential protection requirements' with which the manufacturer must comply. Testing will involve interpreting and adapting the generic standards for Gensets to take account of their physical size. Each manufacturer may interpret these standards differently and therefore across the industry equipment may not be assessed in a 'standard' way. Hence, from a manufacturer's point of view there is no way of knowing whether competitors (or competent bodies) will have prepared TCFs in the same way.

A further complication which must be addressed is whether the equipment is 'placed on the market' or 'taken into service'. Since contracts for Gensets are usually placed as a result of a tender or by the manufacturer being approached by his client, then it is reasonable to assume that the equipment is 'taken into service' and as such does not need to carry the CE marking but it is still required to comply with the protection requirements.

This apparent state of affairs from a Genset manufacturer's viewpoint is unsatisfactory. The sensible approach would be for the manufacturers, using their trade association as a forum, to derive test procedures which could be drawn up as guidelines for use by the industry. These could be fed into the standards making machinery *via* BSI, to be eventually adopted by CENELEC. Such action would hopefully produce 'relevant' standards and therefore remove the current dichotomy outlined above. Alternatively the set of guidelines might enable the generic standards to be interpreted and applied uniformly by the industry within the context of TCFs.

As already indicated a Genset consists of identifiable sub-units and each Genset manufacturer uses the same suppliers for engines and alternators, therefore measurements made on a range of power ratings are likely to be representative of any particular manufacturer's product.

Practical testing methods applied to gensets
Having already determined that relevant standards are not available for assessment of the EM characteristics of Gensets, tests will have to be derived from available relevant and non-relevant standards and also from the experiences of known sources of expertise.

i) Emissions

Conducted Conducted emissions can be measured using a LISN. LISNs with current ratings up to 100 A are available, for larger Gensets a suitable LISN may have to be constructed or the voltage probe defined in EN 55 011[11] (see Chapter 6, Figure 6.8) could be used.

Radiated Radiated emissions should be measured on an Open Field Test Site. For the larger sets a 30 m site will be required. Such testing will require considerable cost for transport of equipment, load banks and commissioning. The alternative is to use a 'makeshift' open field site (car park, *etc*) and move the antennas around the Genset. The distance between Genset and antenna must be at least 10 m. This method is described in EN 55 011[11] and prETS 300 127[12] for emission testing of physically large systems (telephone exchanges).

It is also necessary to relate the EM performance of the constituent parts to the complete system. For example a control cubicle can be characterised for radiated emissions on a 10 m Open Field Test Site. It will be necessary for the control equipment to be exercised over its normal operating conditions *via* suitable dummy signals, this will assess the likely major sources of emissions within the cubicle *viz* the control electronics, the automatic voltage regulator (AVR) *etc* [Parker, 1980[13]]. It may also be necessary to test constituent parts of the control system in isolation, such as the AVR, in order to establish a 'reference test' which could be carried out during normal production testing. Conducted emission measurements can also be made on the sub-assemblies. If sub-assemblies are characterised for a known system performance, then new designs of sub-assembly may be substituted if their EM performance is within the older version's performance envelope, without the need to retest the complete system.

ii) Immunity

Conducted The equipment most likely to produce maloperation or performance degradation of the system is the control electronics. Accordingly if the 'fast transient burst' specified in IEC 801-4[14] is taken as an example, this could be coupled into the cables feeding the control system power supply and the effects observed on the performance of a Genset under test bed conditions. Should the AVR and synchronising circuits be run from individual power supplies then the disturbance can also be coupled to the feeds to these units.

Radiated In order to subject a Genset to an electric field of say 10 V/m [severity level 3, IEC 801-3[15]], it will be necessary to test such equipment in a large anechoic chamber such as the MIRA facility used by the motor industry. However the costs would be prohibitive for the Genset industry where, as already mentioned, margins and production volumes are low. Hence it is necessary to devise a 'reasonable' test methodology. An acceptable method may be to use portable screening and absorbing material to surround the control cubicle and then generate 10 V/m within this much smaller volume. Again the effects on the performance of the Genset can be observed. This sort of test does not show the effect of the engine or skid acting as a large antenna into which even a low field strength may couple, which might induce currents onto sensor cables, *etc*. Hence it may be necessary to test a 'typical' Genset in a chamber and also to subject parts of the system to radiated fields as described above, to establish a correlation, perhaps enabling subsequent radiated immunity tests to be limited to those sub-assemblies which are known to be susceptible *eg*. the control electronics.

ESD The effects of ESD [IEC 801-2[16]] on a system may be established on the test bed where any handles, switches, membrane switches, keypads *etc*., normally touched by an operator can be subjected to either an air discharge or contact discharge as defined in the standard.

A problematic area for immunity testing is the severity level to be tested. This is selected for the EM environment in which the equipment will be operating. Ideally typical Genset installations should be measured to determine the local threat posed to the set by radiated fields and mains borne disturbances. These levels then need to be compared to the immunity measured in the tests outlined, to establish whether further 'hardening' of the control electronics or the cabling to it is required.

From the programme of measurements outlined the emission and immunity characteristics for representative Gensets and their sub-assemblies will have been determined. These will indicate the typical emission levels both radiated and conducted and identify EMI sources (*eg*. control electronics, AVR, rotating diode assemblies *etc*.,)

From site measurements the threat posed by the environment will have been evaluated and will give guidance for acceptable severity levels of radiated field strength and mains borne interference. From these measurements EMC guidelines may be derived for test, design and installation of Gensets.

The problems of demonstrating compliance for large equipment, systems and installations is discussed further in Chapter 11.

10.3.4 Data gathering equipment (industrialised personal computers, dataloggers, *etc.*)

Data gathering equipment is now commonly used in various factory environments. Such equipment usually consists of 'industrialised' personal computers, that is PCs housed in robust enclosures or dedicated microcomputers distributed around a factory. These may be either connected to sensors on a machine or process, or may have data input manually through an integral keyboard. Such microcomputers are usually interconnected or connected to a centralised computer system. Inevitably such equipment and its cabling are subjected to a severe EM environment.

When considering the action plan the microcomputers themselves may be considered as being covered by EN 50 081-1[17] and EN 50 082-1[6] the generic emission and immunity standards for the 'residential, commercial and light industry' environment, or the generic standards for the industrial environment [EN 50 081-2[8], prEN 50 082-2[9]]. Where we can be specific and say that the equipment is being used for industrial process measurement or control, then immunity is covered by HD481[17] (IEC 801). Some caution is needed as the generic industrial immunity standard is still in draft form and HD 481[17] has not appeared in the OJ.

Ideally a manufacturer should carry out an EM site survey to establish the threat to his equipment from radiated fields and conducted interference. From these site measurements the threat posed by the environment can be evaluated which will give guidance for acceptable severity levels of radiated field strength and mains borne interference. (Subsequently these severity levels could be offered to BSI and hence to CENELEC as the basis for an application guidance document to the use of the generic standards.) The environmental threat may also be established by reference to the draft IEC 1000-2-5[10] which characterises EM disturbances for different types of environment.

Caution needs to be exercised as the Data Gathering Equipment manufacturer may be under contract to supply a system to a client, or an installation contractor may purchase the data gathering equipment from a subcontractor/supplier and install a system. Such a system 'taken into service' is within the scope of the EMC Directive, although systems of this type may be difficult to police. Testing of a distributed system is not covered by an EN and therefore the installer must use the TCF route for complete systems. If it can be argued that the system is not a system but an installation, then it is excluded by the UK legislation provided that the constituent apparatus (or systems) are compliant. In this instance it is necessary to demonstrate that the constituent apparatus is not designed by the manufacturer for use as or supplied as a single functional unit.

As discussed for Diesel Generators, a 'reasonable approach' is needed and it is suggested that measurements should be made on a number of typical installations and recorded in the TCF (note under the FCC Rules measurements at 3 sites are required). This in effect will verify the system design and installation procedures and if supported by quality assurance procedures demonstrates to a competent body that other systems installed in the same way and using the same constituent apparatus are also likely to meet the protection requirements.

Since such a system is totally within a factory environment, any interaction with the factory services will be the responsibility of the installer under standard contract procedures. However the effect of such a system beyond the factory boundary needs to be considered and in this respect the data gathering equipment is now similar to ISM equipment. This infers that, for radiated emissions, as long as emissions do not exceed the ISM limits measured from an external wall (see Chapter 6) then the equipment will be deemed satisfactory. For individual units complying with EN 55 022[4] this should not be a problem, irrespective of the lengths of interconnecting cables. This is the view represented in the explanatory document[11] which states that 'each apparatus or system used in the installation is subject to the provisions of the EMC Directive — and that — apparatus and systems must comply with the installation conditions laid down by the manufacturer of the apparatus or system to ensure the proper operation of the installation'.

Summarising:

1) The *installer of data gathering equipment* should:
   i) determine the 'threat' and hence the severity level to which equipment should be tested;

ii) compile measurements made at various sites for inclusion within a TCF and it is suggested that these measurements should include those made outside the building as laid down for ISM equipment (see Chapter 6) [EN 55 011[10]].

2) The *manufacturer of data gathering equipment* should:
i) carry out emission testing to EN 55 022[4] (EN 50 081-1[16]) or EN 55 011[11] (EN 50 081-2[8]);
ii) immunity testing to EN 50 082-1[6] or -2[9], to severity levels specified by the system installer;
iii) prepare detailed installation guidelines for the equipment specifying for example cable types and identifying practices which may affect the emissions from or the immunity of the equipment.

10.4 Summary

In this chapter four examples of products or installed systems have been considered. These have served to illustrate the use of the 'Action Plan', together with an up-to-date list of standards, to determine a route for demonstrating compliance with the EMC Directive. A number of concerns have also been illustrated:

- the concepts of 'placing on the market' and 'taking into service';
- difficulties in interpreting the standards;
- dealing with sub-assemblies;
- adapting test methods for use in the TCF;
- the difficulties and cost implications of testing large and distributed installations;
- and how trade associations can be involved in the standards making process.

References

1. 89/336/EEC Council Directive 'on the approximation of laws of Member States relating to electromagnetic compatibility', Official Journal of the European Communities No.139 25 May 1989, pp19-26
2. 92/31/EEC Council Directive of 28 April 1992 'amending Directive 89/336/EEC on the approximation of laws of the Member States relating to electromagnetic compatibility', Official Journal of the European Communities No L 126/11, 12 May 1992

3. Statutory Instruments 1992 No. 2372 'The Electromagnetic Compatibility Regulations 1994, HMSO, October 1992
4. EN 55 022: 1987 (BS6527:1988) 'Limits and methods of measurement of radio interference characteristics of information technology equipment', British Standards Institution, 1988
5. prEN 55 101 Immunity of Information Technology Equipment; Part 2 (BSI 89/34172 DC) ESD; Part 3 (BSI 89/34171 DC) immunity to radiated electromagnetic energy; Part 4 (BSI 90/30270 DC) immunity against conducted signals; NB.to be renumbered prEN 55 024 Parts 2,3 and 6
6. BS EN 50 082-1: 1992 'Electromagnetic compatibility - generic immunity standard Part 1. Residential, commercial and light industry', British Standards Institution, 1992
7. EC explanatory document on Council Directive 89/336/EEC, 111/4060/91/EN-Rev. 1, 1991
8. BS EN 50 081-2 'Electromagnetic compatibility - generic emission standard generic standard class: industrial', British Standards Institution, 1994
9. prEN 50 082-2: 1991 Generic immunity standard, Industrial (BSI 91/21828 DC)
10. IEC 1000 Part 2 Section 5 'Environment. Classification of electromagnetic environments.' IEC [TC 77(Secretariat)108]
11. BS EN55011: 1991 'Limits and methods of measurement of radio disturbance characteristics of industrial, scientific and medical (ISM) radio-frequency equipment', British Standards Institution, 1991
12. pr ETS 300 127 'Radiated emission testing of physically large systems', European Telecommunications Standards Institute, 1990
13. W H. Parker 'Electromagnetic control of diesel engine generators', IEEE, 1980, CH1538-8/80/0000-0172$00.75 pp 172-176
14. IEC 801-4 'Electromagnetic compatibility for industrial-process measurement and control equipment Part 4: Electrical fast transient/burst requirements', IEC, first edition 1988
15. IEC 801-3 'Electromagnetic compatibility for industrial-process measurement and control equipment Part 3. Radiated electromagnetic field requirements', IEC, 1984
16. IEC 801-2 'Electromagnetic compatibility for industrial-process measurement and control equipment Part 2: Electrostatic discharge requirements', IEC, second edition 1991
17. HD481 CENELEC harmonisation document adopting IEC 801 parts 1, 2 and 3 [14, 15 and 16]

11
EMC in Large Systems

11.1 Introduction

The EMC Directive[1] applies to all electrical or electronics equipment and equally to consumer products and physically large systems. In the case of the large systems however, demonstrating compliance is not so straightforward.

For single apparatus or small systems, standards covering emission and immunity testing have been developed by the IEC and adopted by CENELEC. These standards are product or product family standards and generic standards covering a wide variety of products used within a defined environment. These standards are discussed in Chapters 6 and 7.

Technical Committee 77 of the IEC has developed a series of basic EMC standards which will be referenced by future or amended existing standards. These standards are in the IEC 1000 series and have been adopted by CENELEC as the EN 61000 series. In practice these basic standards incorporate the measurement techniques used by existing product specific CISPR and IEC standards.

There are many difficulties associated with EMC measurements to be made on large equipment, systems or installations

- their physical dimensions
- the lack of relevant standards, and
- in many instances, large systems are never identical.

Examples of large equipment are:

- railway rolling stock
- textile machinery
- structured wiring systems and associated ITE networks
- telephone exchanges, and
- power generation equipment.

For physically large equipments only draft standards are currently available. An example is prETS 300 127[2] (prEN 50098, 91/03506) which has been drafted around the emission requirements for telephone exchanges. Standards are being developed by the railway industry (eg. prEN 50121-3-

2:1994[3]), within the IEC committee structure, which are likely to find applications elsewhere.

11.2 Compliance with the EMC Directive for Large Equipments

Since the publication of the EMC Directive many questions have been raised about how to demonstrate compliance with the protection requirements for large equipment, systems and installations. From Article 1 of the Directive, it is clear that all electrical and electronics apparatus, systems and installations which are liable to cause electromagnetic disturbance or which may have its performance affected by such a disturbance, are within the scope. These questions have been partly clarified by the European Commission's explanatory document[4] which defines apparatus, systems and excluded installations (see Chapter 3), these are repeated here for completeness.

Apparatus or equipment are the same and mean a finished product having an intrinsic function, intended for use by a 'final user' and intended to be 'placed on the market' as a 'single commercial unit'. Examples are a domestic sewing machine or an electric lawnmower.

A *system* means several items of apparatus combined to fulfil a specific objective and intended to be placed on the market as a single functional unit. An example of a small system is a personal computer to which can be added a number of peripherals. Article 3 of the EMC Directive states that equipment should comply 'when it is used for the purposes for which it is intended' and therefore the system must comply.

An *excluded installation* means several items of apparatus or systems put together at a given location to fulfil a specific objective but not intended to be placed on the market as a single functional unit. However each item of apparatus or each system must be individually compliant and carry the *CE marking*.

Despite these definitions there remains debate, particularly regarding unique large systems which have specific functions but are *taken into service* and where the component sub-assemblies are of bespoke manufacture and do not comprise a single functional or commercial unit until the system is installed and commissioned. From the definitions this is not actually an excluded installation, but ***an installed system*** and is therefore required to be compliant. Because it is taken into service and, it may be argued, is not placed on the market, whilst it must be compliant it does not need to carry the CE marking. Conformance will be the responsibility of the installer. In some instances this may actually be the user.

In order to demonstrate compliance without standards it is necessary to use the Technical Construction File or TCF (see the action plan flowchart Figure 10.1).

The Department of Trade and Industry hasissued guidelines[5] on the preparation of TCF's. These detail approaches to different types of equipment including 'one-offs and variants, but also for physically large equipment. The emphasis is placed on the *technical rationale* which is being adopted to demonstrate conformance with the protection requirements of the EMC Directive detailed in Article 4 (see Chapter 2) and restated here:

— *equipment must be constructed to ensure that any electromagnetic disturbance it generates allows radio and telecommunications equipment and other apparatus to function as intended; and*
— *equipment must be constructed with an inherent level of immunity to externally generated electromagnetic disturbances.*

Manufacturers who need to claim compliance for their product by generating a TCF will be required by a Competent Body to show that the product meets the protection requirements of the EMC Directive. This will almost certainly require test results. The DTI recommends that the manufacturer should consult with his chosen Competent Body at the outset. The Competent Body should be able to advise which tests they require to be included within the TCF. More generally the tests which will be required are likely to be those which are prescribed by the generic standards but adapted to suit the particular product under consideration. In addition, because of features of the product or the environment within which it will be used, there may be specific tests required to demonstrate that the product meets the protection requirements.

In order to meet these requirements a typical TCF contents will consist of three parts:

Part I will include an identification of the apparatus/equipment and a description of it.

This is followed in part II by the technical rationale, details of significant EMC design elements and test evidence where this is appropriate.

Part III will contain the report or certificate from a Competent Body.

The manufacturer may initially select a Competent Body by consulting the Competent Body List[6] which provides location and telephone number. More detailed information regarding the types of product which Competent Bodies have experience of, and therefore the TCF content they are 'competent' to assess, may be obtained from the DTI (Fax 0171-215 1529/2909). Discussion with the Competent Body should result in a programme of work enabling the manufacturer to complete Parts I and II of the TCF. The contents of the TCF are given in appendix H. This will almost certainly include a test programme which the manufacturer may carry out himself, sub-contract to a third party or to the Competent body itself.

Prior to discussion with the Competent Body the manufacturer may wish to carry out pre-compliance tests to establish whether his product is likely to comply. It is suggested that the manufacturer attempts to apply or adapt the tests described in the generic standards. Initially the operating environment should be considered in order to establish permissible emissions and identify potential EMI threats to normal operation.

Information is provided on each of the following types of large installation in the DTI guidelines[5]:

- an installed system comprised of compliant equipment/apparatus
- a single large equipment or comprised of large sub-assemblies

Large installed systems comprised of compliant equipment/apparatus (TCF)

The description of the apparatus should include details of the physical location or typical locations within which the equipment will be installed. The characteristics of the location should be described for example the construction of the building in which the installation is housed and significant local environmental factors such as the proximity of sources of EM disturbance. Such information may easily be described for an installation at a specific site, if the TCF is to be used generally for similar installations then the environmental information should cover all likely locations.

Test data will detail the standards with which the equipment comprising the installation conforms and any additional tests performed on the equipment. Details may be included of any measurements made on the installation or typical installations. Both EN 55 022[7] and EN 55 011[8] suggest methods for making emission measurements on large installations which may be adapted for this purpose. Particularly relevant are the emission measurements described which are performed externally to the building enclosing the equipment.

The design practices used to confer EMC to the system should be described (eg signal wiring separated from the power wiring and run in separate conduits or the use of screened cables for signal wiring) along with the installation and maintenance guidelines.

Large single items of equipment or comprised of large sub-assemblies (TCF)

The description of the apparatus should include details of the physical location or typical locations within which the equipment will be installed. The characteristics of the location should be described for example the construction of the building in which the installation is housed and significant local environmental factors such as the proximity of sources of EM disturbance. Such information may easily be described for an installation at a specific site, if the TCF is to be used generally for similar installations then the environmental information should cover all likely locations.

In situ testing may be performed on large equipment either at a manufacturers premises or at an installation or typical installations. Both EN 55 022[7] and EN 55 011[8] suggest methods for making emission measurements on large installations which may be adapted for this purpose. There is also the draft ETSI standard, prETS 300 127[2], which details a radiated emission method based on EN 55 022[7], whereby the antenna is moved around the equipment. Particularly relevant are the emission measurements described which are performed externally to the building enclosing the equipment. It may be possible to perform testing on sub-assemblies and then justify that the complete equipment complies on the same basis as for an installation of compliant apparatus.

The design practices used to confer EMC to the equipment installation should be described (eg signal wiring separated from the power wiring and run in separate conduits or the use of screened cables for signal wiring) along with the installation and maintenance guidelines.

So how do we go about demonstrating compliance for a large equipment using the TCF?

11.3 Test methods for physically large installed systems

Firstly we must consider emission and immunity test methods in the context of large equipment.

11.3.1 Emission testing [Finney[9]]

The draft emission standard for large telecommunications equipment, prETS 300 127[2], is an attempt to apply the principles of emission testing defined by CISPR 22[10] and its European counterpart EN 55 022[7], to physically large systems.

The major limitation in applying EN 55 022[7] to large equipment is the fundamental limitation of transporting equipment to, and making radiated emission measurements on, an Open Field or Open Area Test Site, details of these are given in Chapter 6 and practical test sites in Chapter 8. The Test site is designed to give repeatable results but measurements may be hampered by the presence of ambient signals.

Large telecommunications systems are assembled from independently operating sub-units and no two systems are identical. Each sub-unit will have its own emission characteristics. They may be connected together with a multitude of differing cable types and lengths which will give the system an emission characteristic having a very broad frequency spectrum.

In order to demonstrate compliance a minimum representative system must be defined which is suitable for testing. Such a system is defined as one which includes at least one of each sub-system type which are connected together in a form representing an installed system. Therefore the representative system is the minimum configurable operational system and becomes the minimum requirement for an equipment under test (EUT), to be used during compliance testing. To ensure that the system being tested is a minimal system, a number of additional sub-units should be added during testing to ensure that there is no significant increase in emission level.

The sources of maximum emisson should be identifed by carrying out investigative pre-compliance measurements. This will enable the representative system to be configured to maximise the emissions from it and to avoid, for example, arranging peripheral equipment in a way which would shield noisy sources.

One of the underlying principles is that equipment for use with a Public Switched Telecommunications Network (PSTN) is designed for continuous updating over a period of time, typically up to 20 years after the original system was taken into service. This is also true of many other large installations. Fundamental to the representative system approach is that new modules may replace old modules without the need for retesting of the system. Each module of the original system should be individually characterised in a defined manner, preferably in accordance with an appropriate standard. This characterisation can then be used as a reference for measurements performed on a new module to determine the effect of significant design changes. Provided the electromagnetic performance envelope of the new module is within the the envelope of the original module the need for repeated system level tests can be avoided.

System cabling may well be a major factor affecting the EMC performance of large telecomms systems. A telephone exchange may typically have ten thousand customer pair cables connected to a main distribution frame located in a basement. Cable runs to the 'equipment floor' may well be several hundred metres in length. Therefore to ensure consistency of measured results the test cables should be arranged in a defined configuration and recorded within the test report. If shielded or special cables are required to achieve compliance then this should also be recorded and identical cables used in actual installations.

It may be necessary to make a judgement, based on the measured results of a representative system, as to whether systems will comply when further modules are added. An assessment will have to be made of the characteristic noise sources of the individual modules to determine whether their frequency components will be additive.

prETS 300 127[2] suggests that radiated emissions from a representative system of up to 6 m by 3 m, may be measured on a 30 m by 30 m Open Field Test Site. To avoid rotating the test equipment to find the maximum emissions, the EUT is positioned in the centre of the site and the measuring antenna is moved around it maintaining a perpendicular distance of 10m from the periphery of the EUT.

EN 55 011[8] also allows for *in situ* measurements to be made by moving the antenna around large EUTs. Emission limits are also detailed for measurements at a prescribed distance away from the exterior walls of buildings, Chapter 6. Whilst this standard is specific to Industrial Scientific and Medical (ISM) equipment, such measurement techniques are equally applicable to other types of large industrial equipment and may be used to provide test data for inclusion within a TCF.

Similarly EN 55 022[7] describes *in situ* emission measurements to be made on large ITE installations. This is described in Chapter 6.

11.3.2 Immunity testing [Delaballe[11]]

All of the immunity testing methods in current usage are based on the IEC 801 series of standards. These were originally developed for 'industrial process, measurement and control equipment', but are now being implemented as basic standards in the IEC 1000-4 series. Fundamentally these standards apply to small equipment, alhough all parts make reference to floor standing equipment. When we consider radiated immunity testing, where it is necessary to illuminate an EUT with an electromagnetic field, the essential constraint is that the EUT must be located within a screened enclosure or anechoic chamber. This is discussed in Chapter 7. Most of the available test facilities can accomodate little more than a cabinet having a base area of 1metre by 1 metre and a height of 2m.

So what is the practical approach?

Firstly consider; When is immunity testing required?

- When it is known or suspected that sub-assemblies or component parts are likely to be susceptible when installed,
- or installation practices may affect the intrinsic immunity of the equip ment,

- that the electromagnetic environment is likely to present a severe threat to the operation of the equipment or even to result in damage,
- or there may be a contractual requirement.

In some instances testing may only be possible when the installation is completed.

Secondly both conducted and radiated threats need to be considered.

A major source of conducted disturbances is the mains supply. An installation may be affected by conducted disturbances present on the supply due to:

- switching operations,
- the nature of other equipment connected to the same supply, and
- the loading of the supply.

The internal threats also need to be considered, for example:

- switching motors on and off,
- the close proximity of variable speed drives to programmable logic controllers, and
- relay switching.

A radiated threat can be presented by the presence of:

- high power radio transmitters,
- ISM equipment using radiated coupling, for example RF wood glueing equipment,
- or mobile low power transmitters such as cellular telephones.

HV and MV sub-station switching operations within an industrial plant can produce both conducted transients and impulsive electromagnetic fields.

The IEC 1000-2 series of basic standards are concerned with the Electromagnetic environment and in section 5[12] the environments are classified.

The locations considered are:

- Residential rural;
- Residential urban;
- Commercial;
- Light Industrial;
- Heavy Industrial, power plant and switchyard;
- Traffic area;
- Dedicated telecommunications centre
- and Hospital.

For each location the disturbance levels, rated on a one to five basis, are defined for different media. For example the radiated threat phenomena considered are:

- Low frequency electric and magnetic fields, and
- High frequency threats from oscillatory, pulsed or non-coherent sources.

When preparing TCFs for apparatus and systems not covered by product specific standards, IEC 1000-2-5[12] provides a resource which may be used to determine the severity levels of immunity tests which can be applied to large installations or in many cases to the sub-assemblies which comprise the installation.

Thirdly we need to consider how established test procedures may be adapted to be used with large systems. These are broadly classified into low and high frequency phenomena.

When we consider conducted low frequency phenomena we are effectively constrained to carrying out measurements on sub-assemblies which can be measured within a test laboratory. The measurements may be based on IEC 1000-2-4: 1994[13] which details the requirements for short term supply interruptions and harmonics.

For immunity to power frequency EM fields measurement techniques are given in IEC 1000-4-8[14]. The severity level must be chosen by carefully considering the final installation arrangement, for example:

- the proximity of power transformers,
- HV or MV busbars or cables.

The intensity of the magnetic field is likely to be between 10 Amps per metre, level 3 and 100 Amps per metre, level 5.

In situ measurements may be made using a small multi-turn loop mounted on an insulating pad. The pad allows the coil to be moved parallel to each face of the equipment and at a fixed distance from it. The number of turns, loop diameter and pad thickness must be determined for the magnetic field to be generated and the physical size which can be sensibly used. This is a simple method which can be used for localised testing on installed equipment.

When high frequency phenomena is considered this falls into the categories of:

- electrical fast transient bursts
- damped oscillatory transients
- and radiated electromagnetic fields.

The effects generated by switching inductive loads can be represented by the simulation of fast transient bursts. IEC 801-4: 1988[15] (Chapter 7) and IEC 1000-4-4[16] provide test techniques which can be adapted to large systems. The output from an electrical fast transient burst generator may be connected to the supply lines via a coupling network or alternatively via a capacitive clamp to signal and control lines. Details of carrying out these tests are included in Chapter 7.

These techniques may be applied to sub-assemblies and also used in-situ. For *in situ* measurements the fast transient bursts may be coupled directly onto the supply via 33nF capacitors without disturbing the fixed wiring. The capacitive clamp may be applied to signal or control lines, if however there are physical constraints then conductive foil having an equivalent capacitance may be used, or it may be necessary to use discrete capacitors. It will be necessary to place the fast transient burst generator on a portable reference ground plane.

The standards provide guidance on the test severity level which should be chosen, this will be dependent on the installation factors such as separation of power and control cables and the operating environment.

By generating a damped oscillatory transient and or a surge, several types of disturbance, *eg*. lightning and high current switching may be simulated. Test methods and waveforms are given by: ANSI/IEEE C62.41 1980[17], the draft standards IEC 801-5[18] and the basic standard IEC 1000-4-5[19].

High frequency electromagnetic fields may couple directly into circuits or may induce currents onto the interconnecting cables which are then conducted into the equipment. To simulate these effects it is either necessary to illuminate either fully or partially equipment with an electromagnetic field. Since RF currents may be induced onto cables and conducted into the EUT, this effect may be simulated by injecting RF currents directly into the cables or cable bundles. This can be carried out using a signal generator, RF amplifier and a current probe which can be applied to the cable bundle. It may be necessary to use a second probe to determine the level of the induced current. These techniques are described in IEC 801 parts 3[20] and 6[21] and the basic standards IEC 1000-4 sections 3[22] and 6[23]. IEC 801-3[20] is described in Chapter 7.

Large measurement facilities of the order of several hundred metres in length do exist for carrying out military NEMP testing, however their use is limited to mobile large equipments such as: warships, military aircraft and armoured fighting vehicles. They are unlikely to be useful for commercial equipmment because of the economic factors.

The main difficulty when considering radiated immunity measurements for physically large systems is that unless equipment is tested at the subsystem level it will not be possible to test it within an anechoic chamber or screened room. Therefore the generated electromagnetic field will not be contained and may interfere with other equipment and the measurement system may constitute an illegal radio transmitter. Nevertheless it may be possible to carry out *in situ* measurements on an installation if the following precautions are taken:

- other equipment in the environment which may be affected should be isolated, this may mean carefully choosing the time of day testing is carried out;
- the field strength should be calculated at the boundaries of the installation to ensure that this is within the levels defined in the standards, account may be taken of attenuation provided by the installation housing;

- it may be necessary to use a antenna and receiver system to ensure that the field strength measured at the installation boundaries do not exceed acceptable levels;
- local screening or use of locally placed 'radio' absorbing materials will restrict the transmission of unwanted emissions

If *in situ* radiated immunity measurements are undertaken then partial illumination can be achieved using standard broadband measurement antennas. The illumination will be partial due to the restriction of the antenna beamwidth. An alternative approach may be to use a single wire matched transmission line loaded with its characteristic impedance, typically 500 ohms. This may be used where an installation is composed of a small number of equipment enclosures. Clearly the measurement conditions are far from ideal, the proximity of the floor and ceiling to the installation will affect the characteristic impedance.

11.4 An approach to large installed system compliance

Having considered the general application of EMC test methods to large systems an approach which will lead to generating the information to be contained within a TCF will be considered. Most large systems contain many electrical sub-systems and it is necessary to identify these and determine a strategy which will lead to producing a TCF. The first essential is to develop an *EMC management plan* for the project.

This should be drawn up at the commencement of the project and typically will include:

- an identification of the sources of interference affecting or likely to affect the system;
- a listing of references, for example to legal documents, customer specifications, standards, or in-house specifications;
- the EMC management rationale, for example whole system EMC requirements and the responsibilities of the prime contractor and his suppliers;
- control of suppliers, this may include the requirement for each supplier to produce an EMC plan and to demonstrate compliance;
- 'whole system EMC management', declaring the overall intention of managing EMC by design, identifying particular ares of concern and an EMC Time management plan.

Appended to this document will be EMC design guidelines and practices an example of which is given in Appendix I.

Whilst it will be necessary to perform some EMC measurements on the whole system, initially it is necessary to identify the various electrical sub-systems and determine the procurement policy from suppliers. In this instance a reasonable approach is to task each sub-contractor with providing documentary evidence that his product is compliant with approprate standards. The use of such standards should be agreed with the selected Competent Body who will ultimately assess the TCF.

As indicated by the management plan, each supplier will become responsible for demonstrating that his equipment meets the EMC requirements specified and will submit his test results to the system contractor who will include these within the system TCF. It is then necessary for the system manufacturer to validate from an EMC viewpoint, the installation and wiring techniques he has used. This will be partly by reference to the management plan which lays down the essential EMC working practices, by reference to the Quality assurance procedures for the contractors organisation, but also by some whole system EMC tests.

Immunity testing may be largely impractical and reliance must be placed on the integrity of the immunity testing performed on the individual items of apparatus or systems. Consideration of applying immunity measurements to large installed systems has been covered earlier, 11.3.2. However radiated emission measurements may be performed by moving the measuring antennas around the system as already described. How all systems may be fully exercised during EMC measurements must also be considered.

For the preparation of a TCF for a large installed system a practical approach has been taken which relies on rigorous testing of sub-systems and the verification of installation and design practices by a combination of managing EMC from the outset, quality assurance procedures and whole system emission testing accepting practical limitations. This approach can be seen to fulfil the essential protection requirements of the EMC Directive in relation to effect of the system on the external environment, the need to control the internal EMC within the system and the manufacturers need to satisfy a court that that he has used all due diligence to avoid comitting an offence!

11.5 Summary

This chapter has considered the EMC of large installed systems, particularly:

- the difficulties of demonstrating compliance with the EMC Directive
- the use of the Technical Construction File
- applicable test methods, and
- the need for an EMC management plan

References

1. 89/336/EEC Council Directive 'on the approximation of laws of Member States relating to electromagnetic compatibility', Official Journal of the European Communities No.139 25 May 1989, pp19-26
2. prETS 300 127 'Radiated emission testing of physically large systems ', European Telecommunications Standards Institute, 1990
3. prEN 50121 Part 3 Section 2:1994 'Railway applications. Requirements for rolling stock apparatus', CENELEC, 1994
4. EC explanatory document on Council Directive 89/336/EEC, 111/4060/91/EN-Rev. 1, 1991
5. DTI 'Guidance Document on the preparation of a Technical Construction File as required by EC Directive 89/336', October 1992
6. DTI 'A list of Competent Bodies'
7. EN 55 022:1987 (BS 6527:1988) :'Limits and methods of measurement of radio interference characteristics of information technology equipment', British Standards Institution, 1988
8. BS EN 55 011:1991 'Limits and methods of measurement of radio disturbance characteristics of industrial, scientific and medical (ISM) radio-frequency equipment', British Standards Institution, 1991
9. Finney, A (GPT Ltd) 'EMC Standard for ITE and Telecommunications', 1992 Regional Symposium on EMC 'From a Unified Region to a Unified World' (Cat. No.92THO463-0), Tel-Aviv, Israel, 2-5 Nov. 1992 Sponsors: IEEE; IEE, pT2.4/1-4
10. CISPR 22:'Limits and methods of measurement of radio interference characteristics of information technology equipment', IEC 1985
11. Delaballe, J (Merlin Gerin) 'Immunity testing of large systems', 1992 Regional Symposium on EMC 'From a Unified Region to a Unified World' (Cat. No.92THO463-0), Tel-Aviv, Israel, 2-5 Nov. 1992 Sponsors: IEEE; IEE, pT2.5/1-5
12. IEC 1000 Part 2 Section 5 'Environment. Classification of electromagnetic environments.' IEC [TC 77(Secretariat)108]
13. IEC 1000 Part 2 Section 4:1994: Compatibility levels in industrial plants for low frequency conducted disturbances, IEC, 1994
14. IEC 1000 Part 4 Section 8:1993: Power frequency magnetic field immunity test. Basic EMC Publication, IEC, 1993
15. IEC 801 Part 4 'Electromagnetic Compatibility for industrial-process measurement and control equipment Part 4. Electrical fast transient/burst requirements', IEC, 1988
16. IEC 1000 Part 4 Section 4: 'Fast transients/burst immunity test', IEC
17. ANSI/IEEE STD C62.41:1991 Surge voltages in low-voltage ac power circuits
18. IEC 801 Part 5 'Electromagnetic Compatibility for industrial-process measurement and control equipment Part 5. Surge immunity requirements', IEC
19. IEC 1000 Part 4 Section 5: 'Surges immunity test', IEC
20. IEC 801 Part 3 'Electromagnetic Compatibility for industrial-process measurement and control equipment Part 3. Radiated electromagnetic field requirements', IEC, 1984
21. IEC 801 Part 6 'Electromagnetic Compatibility for industrial-process measurement and control equipment Part 6: Immunity to conducted radio frequency disturbances above 9kHz', IEC
22. IEC 1000 Part 4 Section 3 'Radiated radio-frequency electromagnetic field immunity test', IEC

12

Other Regulations Summarised and Interpreted

This chapter is principally concerned with reviewing the US FCC EMC regulations, the preDirective regulations in Germany, the UK and other EC Member States. It should be read by those intending to market their products under the existing regulations during the transitional period to December 1995 and those who have been required to meet either FCC or VDE requirements in the past. An overview is presented of the FCC 47CFR Parts 15Jand 18, also VDE 0871 and a comparison is made of the emission limits with the harmonised European standards. It is shown that products are likely to meet the requirements of the harmonised standards if they have previously complied with either the US or German regulations.

12.1 Introduction

It is important to review EMC regulations to which UK industry has had to respond to in the past, as this is the 'knowledge' platform it is operating from in trying to meet the demands of the EMC Directive [89/336/EEC[1]]. Also during the transitional period ending 31 December 1995, existing European regulations in each member state will continue to be enforced. The UK EMC regulations were previously implemented by the Wireless Telegraphy Acts[2]. Other EMC regulations to which some sectors of the UK industry have had to respond are:

i) outside Europe the USA Federal Communications Commission, FCC, requirements, when exporting to North America

ii) within Europe the Fernmelde Technisches Zentralamt (FTZ) enforced Verband Deutscher Elektrotechniker (VDE — Association of German Electrical Engineers) standards when exporting to the German

Federal Republic. The role of the FTZ has now been taken by the Zentralamt für Zulassungen im Fernmeldewesen (ZZF — Central Office for Telecommunications).

The German government made significant changes to its laws prior to 30 June 1992 and these are reviewed [EMC Europe, 1995[3]]. By 1996 the VDE EMC standards are required to be harmonised with the appropriate Euro Norm (EN). Where the VDE EMC standards have been considered as a barrier to trade in the past, this will no longer be possible under the rules of the EMC Directive, however compliance with the EMC Directive will of course be required and this is implemented by the EMVG law in January 1996!

12.2 Existing Legislation and EMC Regulations in the UK

Prior to the EMC Regulations (1992), EMC legislation in the UK was an integral part of the Wireless Telegraphy Acts[2], 1949 and 1967. The WT Acts[2] were subsequently amended by the Telecommunications Act 1984[4] (T.Act) [Electrical Interference: A Consultative Document, 1989[5]].

The WT Acts[2] are essentially limited to Wireless Telegraphy or Radiocommunications apparatus used for emitting or receiving EM energy and operating at frequencies less than 3000 GHz. Section 10 of the 1949 Act[2] enables regulations to be made for controlling both radio and non-radio equipment which might cause interference to radio-communications. Section 7, superseded by section 77 of the T.Act[4], gives the Secretary of State power to restrict the sale or use of radio apparatus which may cause undue interference.

Regulations were implemented to control the emissions from some categories of non-radio apparatus *ie*. unintentional transmitters. These are: spark ignition systems, electro-medical apparatus, RF heating devices, household appliances, portable tools, fluorescent lights (luminaires) and CB radios.

Immunity was only considered in respect of radio equipment. Section 12A of the 1949 Act[2], as amended by the T.Act[4], allowed immunity regulations to be implemented, however this provision has not been used.

Radio equipment licensed under the WT Acts[2] must generally comply with MPT specifications. EMC is included within these.

When the provisions of the WT Acts[2] are compared with the 'scope' of the EMC Directive, both in terms of the equipment and the EM phenomena covered, it is apparent that major amendments would have been required to modify the legislation, hence new primary legislation was required. Earlier EC EMC Directives were included by amendment to the W T Acts. These are 76/889/EEC[6] and 76/890/EEC[7]; they cover domestic electrical equipment and luminaires. Both of these Directives and the amendments to them (household appliances: 82/499/EEC, 83/447/EEC and 87/308/EEC; fluorescent lighting apparatus: 82/500/EEC, 83/447/EEC and 87/310/EEC), are repealed by the EMC Regulations[22] from January 1996 which covers all electrical and electronic equipment.

In the UK the Directives on household appliances and luminaires were enforced by amendments to the 1949 Wireless Telegraphy Act[2] and implemented by 'statutory instruments'. For household appliances, portable tools, *etc.*, SI 1978/1267 amended by SI 1985/808 and for fluorescent lighting apparatus, SI 1978/1268 amended by SI 1985/807. These SI's are complied with by testing in accordance with BS 800[8] and BS 5394[9].

Other UK EMC regulations include: SI 1963/1895 Electro-Medical Apparatus, SI 1971/1675 RF Heating Apparatus, SI 1973/1217 Motor Vehicles — interference from ignition apparatus, SI 1984/1216 Reg 43 Passenger Ships and SI 1984/1217 Reg 42 Cargo Ships.

In the UK in addition to the WT regulations, there were guidelines for EMC practice laid down by a number of British Standards (which are not enforced), these were listed in the IEE publication 'Electromagnetic Interference' September 1987, a report of a Public Affairs Board Study Group[10]. These standards are now harmonised or being harmonised with the European Standards (EN) drafted by CENELEC for demonstrating compliance with the EMC Regulations which became effective during 1992 and are fully implemented on 1 January 1996, see Tables 5.2 to 5.5.

12.3 The US FCC Regulations

The FCC administers the use of the frequency spectrum in the USA. Title 47 of the code of Federal Regulations[11] covers Telecommunication and contains, in five volumes, the intentional and incidental use of the spectrum.

The parts relevant to EMC are contained in Chapter 1: Part 15 'Radio Frequency Devices' and Part 18 'Industrial, Scientific and Medical Equipment'. This was revised as of October 1, 1993.

A summary is given here of the salient features of these regulations which are pertinent to our understanding and interpretation of the EMC Directive and which may be pertinent to manufacturers complying with the FCC regulations and wishing to use existing test results to demonstrate compliance with the emission aspects of the EMC Directive via the Technical Construction File (TCF) route. For manufacturers needing to comply with the FCC regulations reference should be to the latest edition of the 'Rules' set out in title 47.

12.3.1 47 CFR Part 15

This 'Part' governs emissions from intentional and unintentional radiators and sets out the regulations, technical specifications, administrative requirements and other conditions to enable equipment to be marketed without an individual license.

The FCC revised part 15 of its rules and regulations in June 1989. This revised set of rules unified many of the limits and extended the measurement range for digital devices or computers up to 5 GHz. In March 1989 the FCC circulated a 'notice of proposed rule making, General Docket No. 89-44, FCC OET TP-5[12]', which proposed measurement procedures and antennas for testing digital devices or computers. Generally these proposals moved the requirements of part 15 closer to CISPR 22 and have been adopted, with some modification, into a revised version of ANSI C63.4-1992[13]. This means that the configuration and the exercising of the EUT should be as for normal use and that the test arrangements should be recorded in the test report, *ie.*, similar to EN 55 022[14] Chapter 6.

12.3.1.1 Subpart A

Part 15 is divided into subparts A, B and C. Subpart A covers general aspects, including definitions, whilst subparts B and C are concerned with unintentional and intentional radiators respectively.

Digital devices

Various definitions of equipment types are given in subpart A, of particular importance are digital devices, previously defined as computing devices. Digital devices are defined as an unintentional radiator generating and using timing signals or pulses at a rate greater than 9 kHz and that use digital techniques, or any device that generates and uses 'radio frequency' energy for performing data processing. This includes telephone equipment. There are two classifications of Digital devices:

Class A A digital device that is marketed for use in a commercial, industrial or business environment.

Class B A digital device marketed for use in a residential environment. This is extended to cover personal computers, calculators and similar electronic devices that are marketed for use by the general public.

Personal computers are specifically considered as Class B digital devices, unless it can be demonstrated that because of price or performance the computer is not suitable for residential or 'hobbyist' use.

Peripheral devices

Peripheral devices are defined as being either external or internal to a digital device. In the latter case when the internal peripheral allows the digital device to be connected by wire or cable to an external device. This definition is extended to cover circuit boards or cards which are designed to be interchangeable internally or externally and which may increase the operating or processing speed of a digital device eg. 'turbo cards'.

Harmful interference

Harmful interference is defined as any emission, radiation or induction that may endanger the functioning of a radio navigation service or other safety services or which seriously degrades, obstructs or repeatedly interrupts a radiocommunications service operating in accordance with the regulations.

Susceptibility

Whilst no technical requirements are specified 'parties responsible for equipment compliance' are advised to consider susceptibility to interference for example the proximity of high power broadcast stations, and to take this into account during the design of the equipment. Equipment suppliers are directed to US Government and non-Government frequency allocations.

Labelling

The labelling requirements are laid down for FM receivers and stand-alone cable input selector switches, all other devices are required to bear the following statement:

This device complies with part 15 of the FCC Rules. Operation is subject to the following two conditions: (1) This device may not cause harmful interference, and (2) this device must accept any interference received, including interference that may cause undesired operation.

Home-built devices and kits

The FCC rules are quite clear regarding home-built devices and kits. Essentially home-built equipment built in quantities of 5 or less is exempt but there is a presumption that the builder will have used good engineering practices. The kit supplier is responsible for demonstating that at least 2 kits assembled in accordance with the instructions meet the rules and is required to supply a label for the builder to attach to the completed unit *(This device can be expected to comply with part 15 of the FCC Rules provided it is assembled in exact accordance with the instructions provided with this kit. Operation etc——).*

Special accessories

The FCC Rules specify that where special accessories *eg* shielded cables or connectors are required to enable either intentional or unintentional radiators to comply with the emission requirements then these must be sold with the equipment. The instruction manual must include 'appropriate instructions on the first page of text' and describe the installation details of how the special accessories are to be used. If special accessories are readily obtained then these do not need to be supplied but must be identified in the instruction manual. The FCC Rules state that: *the resulting system, including any accessories or components marketed with the equipment, must comply with the regulations.*

Inspection by the Commission

Any equipment or device covered by part 15, together with any certificate, notice of registration or any technical data required to be kept on file is to be made available for inspection by a Commission representative upon reasonable request. The Commission from time to time may ask for equipment to be submitted for test at its laboratories. Such testing will be performed using the measurement procedures which were in effect at the time the equipment was verified.

Measurement standards

Measurement procedures are identified in part 15 subpart A, to determine compliance with the technical requirements. An example is FCC/OET MP-4 (1987), the FCC procedure for measuring RF emissions from computing devices. This particular procedure was superseded after 1 May 1994 when a revised verification procedure for digital devices came into effect. This revised procedure also becomes applicable to intentional and other unintentional radiators after 1 June 1995, when equipment should be measured in accordance with American National Standards Institute (ANSI) C63.4-1992, entitled 'Methods of mesurement of radio-noise emissions from low voltage electrical and electronic equipment in the range of 9 kHz to 40 GHz'.

Radiated emission measurements shall be performed on an open field test site (OFTS), but other sites are permitted if they are calibrated so that the measurement results correspond to what would be obtained on an OFTS. The FCC have announced that they will accept results from a GTEM (see appendix F).

Where only *in situ* measurements are possible, measurements must be performed 'at a minimum of three installations that can be demonstrated to be representative of typical installation sites.'

Guidance is given in the Rules on variation of testing distance and it is stated that measurements shall be performed at a sufficient number of radials around the EUT to determine the radial at which the radiated emissions are at a maximum.

For composite systems devices shall be tested with the accessories attached and the device under test fully exercised with these external accessories. Again the system should be configured to maximise the emissions. Only one test using peripherals or external accessories that are representative of the devices which will be used with the EUT is required. Where multiple accessory ports exist, an external accessory must be connected to one of each port type. *'All possible equipment combinations do not need to be tested.'* The measured emissions of the composite system must not exceed the highest level permitted for an individual subsystem. For a combination of Class A and Class B devices intended for use in a Class A environment, the composite system is only required to meet the Class A limits. However if the central control unit is Class B then it must be demonstrated that it continues to meet Class B with the Class A internal peripherals installed but inactive.

Frequency range of radiated measurements

Broadly the frequency range for radiated emissions for both intentional and unintentional radiators is given in Table 12.1. Full details are given in the Rules and variations for non digital device unintentional radiators.

Table 12.1 FCC Part 15 upper frequency limits

EUT Highest internally generated frequency	Frequency range of measurements
1.705 - 108 MHz	30 - 1000 MHz
108 - 500 MHz	30 - 2000 MHz
500 - 1000 MHz	30 - 5000 MHz
Above 1000 MHz	5th harmonic of the highest frequency or 40 GHz, whichever is lower

12.3.1.2 Subpart B - Unintentional radiators

Emission limits are laid down for the two classes of digital device along with measurement procedures and compliance requirements.

Certain 'digital devices' are excluded from subpart B, including:

a) those used in transportation vehicles (motor vehicles and aircraft);
b) those used exclusively as an electronic control or power system utilised by a public utility or in an industrial plant;
c) those used exclusively as industrial, commercial or medical test equipment;
d) those utilised in an (domestic) appliance;
e) specialised medical equipment used for patient treatment under the guidance of a medical practitioner either in the home or at a health care facility;
f) digital devices with a power consumption of less than 6 nW;
g) joystick controllers or similar devices containing non-digital circuitry;
h) devices where the highest frequency used or generated is less than 1.705 MHz which cannot be operated from ac power lines. Note

digital devices which operate from battery eliminators or chargers or ac adaptors while connected to the mains are not exempted;

i) composite equipments are only exempt if all of the devices employed qualify for exemption.

Although not mandatory, manufacturers of such equipment are strongly advised to meet the limits.

Compliance

Class A *Verification* tests for Class A digital devices must be performed by the manufacturer and test results kept on file. Certification by the FCC is not required.

Class B Personal computers and peripherals are required to be *certificated* by the FCC. This is done by examining the manufacturer's test results.

Other Class B equipment must be *verified* as for Class A devices. These include calculators, digital clocks, and watches and electronic games.

For Class A digital devices the following statement must have a prominent location in the instruction manual:

NOTE: This equipment has been tested and found to comply with the limits for a Class A digital device, pursuant to part 15 of the FCC Rules. These limits are designed to provide reasonable protection against harmful interference when the equipment is operated in a commercial environment. This equipment, generates, uses and can radiate radio frequency energy and, if not installed and used in accordance with the instruction manual, may cause harmful interference to radio communcations. Operation of this equipment in a residential area is likely to cause harmful interference in which case the user will be required to correct the interference at his own expense.

For Class B digital devices the following statement must have a prominent location in the instruction manual:

NOTE: This equipment has been tested and found to comply with the limits for a Class B digital device, pursuant to part 15 of the FCC Rules. These limits are designed to provide reasonable protection against harmful interference in a residential installation. This equipment, generates, uses and can radiate radio frequency energy and, if not installed and used in accordance with the instruction manual, may cause harmful interference to radio communications. However there is no guarantee that interference will not occur in a particular installation. if this equipment does cause harmful interference to radio or television reception, which can be determined by turning the equipment off and on, the user is encouraged to try to correct the interference by one or more of the following measures:

Reorient or relocate the receiving antenna.

- *Increase the separation between the equipment and receiver.*
- *Connect the equipment into an outlet on a circuit different from that to which the receiver is connected.*
- *Consult the dealer or an experienced radio/TV technician for help.*

Limits

The FCC has established limits for radiated and conducted interference for both A and B device classifications. The limits for Class B are more stringent as residential users are less likely to be able to cope with 'harmful interference'. Measurements should be made using a quasi-peak detector receiver.

Conducted emissions

For either a Class A or Class B digital device designed to be connected to the public utility ac mains, the conducted emission limits are given in Table 12.2.

Table 12.2 Conducted Limits as defined in 47 CFR Part 15, Subpart B

Frequency MHz	Mains RF Line Voltage μV	
	Class A	Class B
0.45 - 1.705	1000	250
1.705 - 30.0	3000	250

These limits may also be expressed in dB referred to 1 dBμV, shown as follows in Table 12.3:

Table 12.3 Conducted Limits 47 CFR Part 15, Subpart B in dBμV

Frequency MHz	Mains RF Line Voltage dBμV	
	Class A	Class B
0.45 - 1.705	60.0	48.0
1.705 - 30.0	70.0	48.0

Radiated emissions

For Class A or Class B digital devices the radiated emission limits are given in Table 12.4.

Table 12.4 Radiated Limits as defined in 47 CFR Part 15, Subpart B

Frequency (F) MHz	Distance m		Field Strength mV/m	
	Class A	*Class B*	*Class A*	*Class B*
30 - 88	10	3	90	100
88 - 216	10	3	150	150
216 - 960	10	3	210	200
960 - 5000*	10	3	300	500

* Depends on clock frequency of device, see Table 12.1

Measurements may be made at closer distances than those specified provided the EUT is in the far field and with the limits increased proportionally by 20 dB per decade.

For ease of use and to enable comparison with other standards it is easier to consider Field Strength in dBμV/m, hence the Table 12.4 is modified for a 3 m test site as follows:

Field Strength in dBμV/m = 20 $\log_{10}$ Ed(μV) referred to 1μV/m.

Table 12.5 ***Radiated Limits 47 CFR Part 15, Subpart B in dBμV/m***

Frequency	**Distance m**		**Field Strength dBμV/m**	
MHz	*A*	*B*	*Class A*	*Class B*
30 - 88	10	3	40.0	40.0
88 - 216	10	3	43.0	43.0
216 - 960	10	3	46.0	46.0
960 - 5000*	10	3	50.0	54.0

* upper frequency determined from Table 12.1

As an alternative to the FCC defined limits, digital devices may be shown to meet both the conducted and radiated emission requirements of CISPR 22, *see* Chapter 6.

12.3.1.3 Subpart C - Intentional radiators

The unintentional emissions from intentional emitters are generally defined to be the same as for Class B digital devices. This subpart details restricted bandwidths and rules for specific type of equipment such as: tunnel radio systems, cable locating equipment and cordless telephones. For further information the FCC Rules[11] should be consulted.

12.3.2 47 CFR Part 18 industrial, scientific and medical equipment (ISM)

ISM equipment is defined as being designed to generate and locally use RF energy at frequencies of 9 kHz and above for industrial, scientific, medical, domestic or similar purposes, excluding applications in the field of telecommunications. Typical applications are listed as the production of physical, biological or chemical effects such as heating, ionisation of gases, mechanical vibrations, hair removal(!), and acceleration of charged particles. Sub-categories of equipment are listed as industrial heating, medical diathermy, ultrasonic, consumer ISM (including microwave ovens, *etc.*).

Upon the request of the FCC the manufacturer, owner or operator of any ISM equipment shall make the equipment and information to indicate that it complies with the requirements, available for inspection.

Note: Imported equipment must be accompanied by a copy of FCC Form 740 certifying that the appropriate equipment authorisation has been issued by the FCC or that the device does not require authorisation and that the device complies with the applicable FCC technical specifications.

Compliance

Consumer ISM equipment is subject to certification prior to use or marketing. Applications for certification (FCC Form 731) must be accompanied by a description of the measurement facilities and a technical report. Consumer ultrasonic equipment of < 500 W and operating below 90 kHz and non-consumer ISM equipment is subject to verification only.

Operating frequencies

ISM equipment may be operated on any frequency above 9 kHz except the frequency bands used for safety, search and rescue: 490-510 kHz, 2170-2194 kHz, 8354-8374 kHz, 121.4 - 121.6 MHz, 156.7 - 156.9 MHz and 242.8 - 243.2 MHz.

The allocated frequencies are shown in Table 12.6.

Table 12.6 FCC and VDE defined ISM frequencies

ISM Frequency	*FCC 'Tolerance'+/-*	*VDE 'Permitted Deviation*
13.56 MHz		0.05%
27.12 MHz	163.0 kHz	0.60%
40.68 MHz	20.0 kHz	0.05%
433.92 MHz		0.20%
915.0 MHz	13.0 MHz	
2450.0 MHz	50.0 MHz	
5800.0 MHz	75.0 MHz	
24.125 MHz	125.0 MHz	
61.25 GHz	250.0 MHz	
122.50 GHz	500.0 MHz	
245.00 GHz	1.0 GHz	

Radiated emission limits

ISM equipment operating on one of the allocated frequencies is permitted 'unlimited radiated energy in the band specified for that frequency'. The radiated emissions lying outside those bands are restricted. As each category of equipment has different specifications only one example will be given here:

Table 12.7 FCC Part 18 radiated emission limits for diathermy equipment

Equipment	**Medical Diathermy**	
Operating at	*Any ISM frequency*	*Any non-ISM frequency*
RF Power, watts	Any	Any
Field strength, µV/m	25	15
Distance, metres	300	300

Note: These field strengths correspond to 68 dBµV/m and 63.5 dBµV/m at 3 m.

Conducted emission limits

Limits are only specified for ultrasonic equipment, inductive cooking ranges and RF lighting devices. The use of a line impedance stabilisation network is specified for making these measurements. Considering RF lighting devices as an example:

Table 12.8 FCC Part 18 conducted emission limits for RF lighting equipment

Frequency MHz	**Mains RF Voltage µV**	
	Non-consumer Equipment	**Consumer Equipment**
0.45 - 1.6	1000	250
1.6 - 30	3000	250

These can be seen to correspond with part 15, Class A and B limits.

12.4 The German EMC Regulations and the VDE Specifications (Verband Deutscher Electrotechniker)

The legal regulations for interference control in Germany are enforced by the ZZF. The standards are developed by the VDE, the Association of German Electrical Engineers in co-ordination with Din Deutsches für Normung (DIN), the German Institute for Standardisation. Standards produced by the VDE which are also adopted as National Standards are referenced DIN-VDE. Standards prefixed 'E' are draft standards.

The corresponding German law to the UK's W T Act of 1949 is the High Frequency Law (HFrG), also of 1949. This law was revised in 1968 and together with revised administrative procedures was published by the German Post Office (BZT) in Verfugung (Vfg.) 523 in 1969. This has been subsequently altered to include the provisions of the Directives concerned with the radio interference from household appliances and fluorescent lights [EMC Europe, 1995[3]].

The principal regulations are:

Vfg. 1044 (1984) — this transposes into German law Directives 76/89/EEC (82/499/EEC) and 76/890/EEC (82/500/EEC) which cover the radio interference requirements respectively of household electrical appliances, *etc* and fluorescent lighting (*see also* 12.2).

Vfg. 1045 (1984) — this regulation describes the procedure for obtaining a General Licence for the equipment covered by Vfg. 1044. In this case 'an appropriate approval mark' is used. Proof of compliance would normally be demonstrated by testing equipment to VDE 0875-1 (domestic appliances EN 55 014), VDE 0875-2 (fluorescent luminaires EN 55 015) or VDE 0875-3 (other equipment particularly semi-conductor controls and systems over 16A).

Vfg. 1046 (1984) — General Licence requirements for radio interference from: data processing equipment and systems, switched mode power supplies, ultrasound equipment, induction and microwave cooking equipment, microprocessor control and equipment employing RF generators, signal generators and electrical measurement and test equipment. The certification procedure is described in section 12.4.1, essentially the equipment defined will be issued with a general licence if it conforms to VDE 0871: 1978 Class B limits. Equipment may be marked with the radio protection mark of the VDE and the index 871B or be provided with a certificate issued by the manufacturer giving details of compliance with the requirements.

Vfg. 242 (1991) — General licence. This was originally intended to replace Vfg. 1045. The changes were limited to essentially the use of voluntary marks and to increasing the scope of the equipment covered to include devices with up to a 25 A current rating. This document was amended by Vfg. 46: 1992 and is now an alternative to Vfg. 1045 until 31 December 1995.

Vfg. 243 (1991) — General Licence. This regulation was intended to replace Vfg. 1046. The certification procedure is again as described in 12.4.1. This document is also amended by Vfg. 46: 1992 and is now an alternative to Vfg. 1046 until 31 December 1995.

A list of principal VDE EMC specifications is given in Appendix C.

12.4.1 VDE 0871: 1978

VDE 0871: 1978[15] is concerned with emissions from industrial, scientific and medical equipment (ISM) that generates or utilises discrete frequencies or repetition frequencies above 10 kHz, it is based on CISPR 11: 1975[16] By virtue of the requirements of Vfg 1046, data processing equipment (and other equipment identified in 12.4) may be granted a General Licence by satisfying VDE 0871: 1978 Class B limits. Therefore the scope of VDE 0871 is broadly the same as for the FCC regulations Parts 15 and 18 and may be compared to the requirements of EN 55 011[17] and EN 55 022[14] already discussed in Chapter 6.

There are three types of licence available to operate equipment generating or using electromagnetic energy at frequencies greater than 10 kHz.

1. A *'General Licence'* — may be granted for a specific type of equipment if the product conforms to *VDE limit B,* equipment may be marked with the radio protection mark of the VDE with the index 871B or be provided with a certificate issued by the manufacturer giving details of compliance with the requirements. The manufacturer also informs the German Post Office (BZT) that the equipment has been placed on the market and must be in a position to provide proof of conformity if required by the BZT. Full details will be published in the Vfg. together with the German text for the certificate.
2. Products only meeting *limit A* require an *individual licence* for each end user. The procedure is set out in Vfg. 523: 1969. The manufacturer applies to the VDE for testing and will be provided with the requirements for conformity (a type test report to the specified limits). This is sent by the manufacturer to the BZT, together with other data about the equipment. The user of the equipment notifies the local Bundespost department which is required to verify that the requirements have been met and then issues the operator with an individual licence.
3. An individual licence may also be granted for specific installations tested *on site* where the relevant requirements of *Class C* may be appropriate.

To avoid the inconvenience of end users having to take responsibility for equipment only meeting Class A limits, manufacturers prefer to test to the more stringent B limit.

As with the FCC regulations, VDE 0871 lays down the ISM frequencies for unlimited radiation, these are shown in Table 12.6 where they can be compared with the FCC frequencies.

Limits

VDE 0871 lays down limits for radiated and conducted interference. These limits are classified A, B and C as referred to in the three types of licence.

Radiated emission limits

Radiated emission measurements are performed in accordance with VDE 0876[18] and VDE 0877[19]. Essentially these requirements are as previously described for EN 55 022[14] and EN 55 011[17]. The RFI field strength limits specified by VDE 0871 are summarised below:

Table 12.9 ***Radiated emission limits VDE 0871: 1978***

Frequency range *MHz*	**RFI field strength** *Open site* *μV/m*				**RFI field strength** *Operational site* *μV/m*		
	Class A		*Class B*		*Class C*		
	30 m	*100 m*	*10 m*	*30 m*	*30 m*	*100 m*	*300 m*
*0.01 - 30		50		50		50	200
**30 - 1000	30		50		30		200
***1000 - 18000	57 dBpw		57 dBpw			57 dBpw	

* Magnetic component only

** This table only shows the toughest limits, for full details see VDE 0871: 1978.

*** These limits apply only to microwave equipment for heating and medical equipment

Table 12.10 ***Radiated emission limits VDE 0871: 1978, expressed in dBμV/m***

Frequency range *MHz*	**RFI field strength** *Open site* *dBμV/m*				**RFI field strength** *Operational site* *dBμV/m*		
	Class A		*Class B*		*Class C*		
	30 m	*100 m*	*10 m*	*30 m*	*30 m*	*100 m*	*300 m*
*0.01 - 30		34		34		34	46
**30 - 1000	30		34		30		46
***1000 - 18000	57 dBpw		57 dBpw			57 dBpw	

* Magnetic component only

** This table only shows the toughest limits, for full details see VDE 0871: 1978.

*** These limits apply only to microwave equipment for heating and medical equipment

Conducted RFI limits

The limits for conducted RFI voltages measured using a LISN [VDE 0876[18]]. Limits are specified from 10 kHz to 30 MHz, as shown in Table 12.12. Essentially above 500 KHz the A and C limit is 60 dBμV, and for

Class B 48 dBμV - it should be noted that these correspond with the FCC Class B limit.

Table 12.11	***VDE 0871: 1978 Conducted emission limits***	
Frequency range	**Limits**	
MHz	Class A Quasi-Peak dBμV	Class B Quasi Peak dBμV
0.01 - 0.15	Decreasing linearly with log of frequency from 91.0 - 69.5	Decreasing linearly with log of frequency from 79.0 - 57.5
0.15 - 0.5	66	54
0.5 - 30	60	48

12.4.2 VDE standards development and the 1992 German 'transitional period' EMC regulations

VDE 0871 has undergone a number of revisions and title changes, the 1978 version was also known as DIN 57871, the 1985 draft version as DIN VDE 0871: 1985 and in 1987 the draft DIN VDE 0871-11: 1987 was issued in line with prEN 55 012. Therefore the harmonised EN 55 011 will be DIN VDE 0871-12.

In 1989 a separate standard to cover data processing equipment was issued harmonised with EN 55 022: 1987[14], DIN VDE 0878-3: 1989. A draft national supplement has been issued to this standard VDE 0878-30[20]. It is the content of this national standard which has been adopted by the German EMC regulation Vfg. 243[21], December 1991, which is of particular concern to manufacturers of equipment covered by EN 55 011[17](VDE 0871-11) or EN 55 022[14] (VDE 0878-3: 1989). Effectively it reinstates the VDE 0871: 1978[15] conducted interference requirements, by extending the limits down in frequency to 10 kHz [VDE 0878-30, Table 1 Class A and Table 2 Class B[20]], it also allows in the case of discontinuous conducted interference the less rigorous requirements of VDE 0875 (EN 55 014, Chapter 6) to be used. More significantly when compared to either EN 55 011[17] or EN 55 022[14] the radiated emission limits are extended both up and down in frequency. For radiated emissions in the range 1 GHz to 18 GHz a limit of 57 dBpw is specified for both Class A and Class B equipment, but a limit of 45dBpw is recommended for Class B [VDE 0878-30[20]], as in the 1978 version of VDE 0871. For the frequency range 9 kHz to 30 MHz, VDE 0871: 1978[15] recommended magnetic field strength limits in terms of mV/m (Table 12.9), Vfg. 243[21] specifies these low frequency limits in terms of magnetic field strength as follows:

Table 12.12	*Vfg. 243 Magnetic emission limits*
Frequency range	**Field strength limit at 3 m, dBµA/m**
9 kHz - 70 kHz	68
70 kHz - 150 kHz	68 - 38 falling linearly with the log of 'f'
150 kHz - 2 MHz	38
2 MHz - 3.95 MHz	38 - 26 falling linearly with the log of 'f'
3.95 MHz - 5 MHz	26 - 22 falling linearly with the log of 'f'
5 MHz - 16 MHz	22 - 2 falling linearly with the log of 'f'
16 MHz - 30 MHz	2

Vfg. 241: 1991 (amended by Vfg. 46) governs the transitional period until the EMVG law (the law implementing the EMC Directive) comes fully into force, this states that until December 1995 manufacturers will have a choice of complying with regulations Vfg 1045 or Vfg 1046 depending on the type of equipment, *or* they may use Vfg. 242 or Vfg. 243, *or* manufacturers may choose to place a product on the market in accordance with the EMC Directive 89/336/EEC[1].

The laws, Vfg 242 and 243, implement the harmonised standards but with additions such as the national standard VDE 0878-30, therefore effectively they provide the same limits as the earlier laws. Therefore to comply with the pre-Directive German regulations, is far more onerous for manufacturers than complying with the harmonised standards which allow compliance with the EMC Directive. Two concerns are apparent therefore:

i) the German representatives involved with CENELEC will presumably try and influence development of the harmonised emission standards to cover the additional frequency ranges which they consider essential for their national standards and regulations and which it could be claimed are required in order to demonstrate compliance with the 'essential protection requirements' of the EMC Directive.

ii) for manufacturers choosing the less onerous option of complying with the EMC Directive by using the harmonised standards, there may be a potential risk that the German authorities may claim that they do not meet the 'essential protection requirements' because compliance has not been demonstrated for requirements which they consider essential nationally. This would of course have to be demonstrated.

12.5 Comparison of FCC/VDE Requirements with the EMC Directive

If the requirements of the FCC and VDE are compared with the EMC Directive and the relevant standards for claiming compliance with it, a number of conclusions can be drawn:

i) the FCC/VDE requirements only cover emissions

ii) the conducted and radiated emission limits in the frequency ranges 0.15-30 and 30-1000 MHz set by the FCC and VDE are very similar, these are also very similar to EN 55 022[14]. Hence equipment complying with the FCC and VDE requirements is also likely to meet the EN emission limits. If we consider the conducted limits at 500 kHz the FCC and VDE limits are the same, 48 dBmV the EN is higher at 56 dbmV. The VDE standard goes down in frequency to 10 kHz whilst the EN and FCC lower frequency limits are 150 kHz and 450 kHz respectively. When the radiated limits are considered then the EN and FCC requirements are similar for example at 30 MHz and at 3 m both limits are 40 dbmV/m, at 100 Mhz the EN is still 40 dBmV/m whilst the FCC requirement is 43 dBmV/m and at 1 GHz the limits are 47 and 46 dBmV/m respectively. This may be considered as 'good news' for manufacturers of equipment already meeting FCC or VDE limits. The FCC permits compliance with CISPR 22 (EN 55 022) as an alternative to the FCC Rules limits, hence any equipment certified or verified using CISPR 22 will automatically meet the requirements of EN 55 022 and hence compliance with the emission requirements of the EMC Directive will have been demonstrated. For a comparison of limits see Table 12.13.

ii) significantly the current German EMC regulations are more strict than those effectively implemented by the EMC Directive, therefore it should be advantageous for manufacturers exporting to Germany to comply with the EMC Directive and apply the CE marking. The question arises however as to how the Germans intend to police the Directive and whether they are able to claim that compliance with magnetic field limits, for example, is necessary to demonstrate compliance with the 'essential protection requirements'.

iii) immunity is not considered by either the FCC regulations (that is quantitatively) or the pre-Directive German regulations, hence all manufacturers are in the same position regardless of whether they have been required to meet FCC or VDE limits. Indeed this is of concern to some US companies who wish to export to Europe as the immunity requirement may become a trade barrier to them. There is also a growing realisation that the European Market Place will be larger than the North American as the relative populations are aproximately 320 million (not including the Eastern European countries) and 250 million respectively!

iv) Undoubtedly the EMC Directive has had an effect on the FCC, as compliance with ANSI C63.4-1992 will be the method of demonstrating compliance for all unintentional emitters from 1 June 1995 and this move starts to divorce the technical requirements from the regulations, just as for the EMC Directive. Further the FCC Rules now expect equipment to have an inherent level of immunity although not quantitatively specified.

Table 12.13 Approximate comparison of EN, VDE and FCC emission limits

Standard	Limit Class	Frequency Range, MHz Conducted Emissions dBµV							Radiated Emissions dBµV/m		
		0.01	0.15		0.5		5	30	230	1000	
		QP	QP	AV	QP	AV	QP	AV	QP	QP	
EN 55022 /	B		66-56	56-46	56	46	60	50	30	37	@ 10 m
EN 55011	A		79	66	79	66	73	60	30	37	@ 30 m
VDE 0871:	B	79-57.5	54		48		48		34	46	@ 10 m
1978	A	91-69.5	66		60		60		30	54	@ 30 m
Magnetic FS	B	34	34		34		34				@ 30 m
dBµV/m	A	34	34		34		34				@ 100 m
FCC Part 15	B				48		48		40	46	@ 3 m
	A				60		70		40	46	@ 10 m

12.6 Pre-Directive Regulations in Other European Countries

UK and German regulations have been described which are enforced in parallel with the EMC Directive until 31 December 1995, under the transitional arrangements. This is also true for the other Member States.

The pre-Directive EMC regulations for member states other than the UK or Germany, are similar to the UK provisions. All States have laws covering radio interference which are enforced by complaint driven procedures. For this reason compliance with either national or European emission standards is usually sufficient under existing regulations. The regulations implement the earlier Directives covering household appliances, fluorescent lighting and spark ignition systems in motor vehicles or agricultural tractors. In addition the regulations are likely to cover the use of high frequency equipment (*ie.*, ISM) and in some Member States the immunity of radio and television receivers.

Further details can be found in 'Electromagnetic compatibility Europe', BSI Technical Help to Exporters, update 6 1994, ISBN 0 580 20917 2[3].

References

1. Council Directive of 3 May 1989 on the approximation of the laws of the Member States relating to electromagnetic compatibility (EMC), 89/336/EEC, OJ L139 of 23.05.89, pp 19-26

2. Wireless Telegraphy Acts 1949 and 1967
3. Technical Help to Exporters 'Electromagnetic Compatibility Europe', BSI, Update 6 1994, ISBN 0 580 20917 2
4. Telecommunications Act 1984
5. Electrical Interference: a Consultative Document', DTI/PUB207/10k 10.89
6. 76/889/EEC 'Radio interference suppression household electrical appliances and portable tools', Council Directive, 1976
7. 76/890/EEC 'Radio interference suppression fluorescent lamps and luminaires', Council Directive, 1976
8. BS 800: 1988 (EN 55 014: 1987) and AMD 6275 (No. 1 effective June 1990) Limits and methods of measurement of radio interference characteristics of household electrical appliances, portable tools and similar electrical apparatus', British Standards Institution, 1988
9. BS 5394: 1988 (EN 55 013) 'Specification for limits and methods of measurement of radio interference characteristics of fluorescent lamps and luminaires', BSI, 1988
10. Electromagnetic Interference', Sept.1987, IEE Public Affairs Board Study Group
12. FCC, 'Code of Federal Regulations No. 47 parts 0 to 19 Telecommunication', Office of the Federal Register National Archives and Records Administration, 1993
12. GEN. Docket No. 89-44 'FCC Procedure for Measuring Electromagnetic Emissions from Digital Devices. FCC OET TP-5 (Draft)', Federal Communications Commission, 1989
13. ANSI C63.4-1992, American National Standard Methods of Measurement of Radio-Noise Emissions from Low Voltage Electrical and Electronic Equipment in the Range 9kHz to 40GHz.
14. EN 55 022: 1987 (BS 6527:1988) 'Limits and methods of measurement of radio interference characteristics of information technology equipment', British Standards Institute, 1988
15. VDE 0871/6.78 (DIN 57 871) 'VDE Specification Radio Frequency Interference Suppression of Radio Frequency Equipment for Industrial, Scientific and Medical (ISM) and Similar Purposes', English Translation by EMACO EMC Consultants
16. CISPR Publication No. 11 'Limits and methods of measurement of radio interference characteristics of industrial, scientific and medical (ISM) radio-frequency equipment (excluding surgical diathermy apparatus)', IEC, 1975
17. BS EN 55 011: 1991 'Limits and methods of measurement of radio disturbance characteristics of industrial, scientific and medical (ISM) radio-frequency equipment', British Standards Institution, 1991
18. VDE 0876 (DIN 57 876) 'Equipment for the Measurement of Radio Frequency Interference, Radio Frequency Interference Measurement Equipment with Quasi-Peak Indicator and Accessories', English Translation by EMACO EMC Consultants
19. VDE 0877 part 2 'Measurement of radio interference Measurement of radio interference field strength', Feb. 1985
20. DIN VDE 0878 Part 30 Draft November 1989 'Electromagnetic Compatibility of Information Technology and Telecommunications Equipment — Limits and methods of measurement of radio interference characteristics of information technology equipment', Translation — Technical Help to Exporters, BSI
21. VFG 243 December 1991 'Radio Interference Suppression of Radio-Frequency Equipment for Industrial, Scientific, Medical (ISM) and Similar Purposes and Equipment used in Information Processing Systems; General Licence', Translation — Technical Help to Exporters, BSI
22. Statutory Instruments 1992 No. 2372 'The Electromagnetic Compatibility Regulations 1994, HMSO, October 1992

Appendix A

List of Harmonised Standards Published in the OJ

Standards which may be used to demonstrate compliance with the EMC Directive must be published in the Official Journal of the European Communities (OJ). Up to April 1995 standards referenced in the OJ were as follows:

EN 50 065-1: 1991

Signalling on low-voltage electrical installations in the frequency range 3 to 148.5 kHz and AM1 1992

Part 1: General requirements, frequency bands and electromagnetic disturbances.

EN 50 081-1: 1992

Electromagnetic compatibility generic emission standard

Part 1: residential, commercial and light industry.

EN 50081-2: 1993

Electromagnetic compatibility generic emission standard

Part 2: industrial environment.

EN 50 082-1: 1992

Electromagnetic compatibility generic immunity standard

Part 1: residential, commercial and light industry

EN 55 011: 1991

Limits and methods of measurement of radio disturbance characteristics of industrial, scientific and medical (ISM) radio-frequency equipment (CISPR 11: 1990 ed 2).

EN 55 013: 1990

Limits and methods of measurement of radio disturbance characteristics of sound and TV broadcast receivers and associated equipment. AM11 1993 (CISPR 13: 1975 ed 1 + Amdt 1: 1983).

EN 55 014: 1987

Limits and methods of measurement of radio interference characteristics of household electrical appliances, portable tools and similar electrical apparatus (CISPR 14: 1993 ed 3).

EN 55 014: 1993

Limits and methods of measurement of radio disturbance characteristics of electrical motor operated and thermal appliances for household and similar purposes, electrical appliances, portable tools and similar electrical apparatus (CISPR 14: 1993 ed 3).

EN 55 015: 1987

Limits and methods of measurement of radio interference characteristics of fluorescent lamps and luminaires (CISPR 15)

EN 55 015 1993

Limits and methods of measurement of radio disturbance characteristics of electrical lighting and similar equipment. (CISPR 15: 1992 ed 4.)

EN 55 020: 1988

Immunity from radio interference of broadcast receivers and associated equipment (CISPR 20 : 1990 ed 2 + Amdt 1 1990).

EN 55 022: 1987

Limits and methods of measurement of radio interference characteristics of information technology equipment (CISPR 22: 1985 ed 1).

EN 60 555-2: 1987
Disturbances in supply systems caused by household appliances and similar equipment
Part 2: Harmonics (IEC 555-2: 1982 ed 1 + Amdt 1: 1985).
EN 60 555-3: 1987
Disturbances in supply systems caused by household appliances and similar equipment
Part 3: Voltage fluctuations (IEC 555-3: 1982 ed 1).
Note: *an EN is an agreed technical text which must be implemented as a National Standard. The EN usually exists as a document referencing an IEC or CISPR standard eg., CISPR 22 and a CENELEC document detailing any agreed deviations. The national harmonised standard is normally the complete text for the EN derived from these two documents eg., BS 6527. It is therefore better to be able to specify an EN as a National equivalent standard to ensure receiving the full text; cross references are provided in Chapter 5, section 5.7.*

Appendix B

List of BSI Publications Relating to EMC

BS 613: 1977
Specification for components and filter units for electromagnetic interference suppression. Last amended 1980.
BS 727: 1983
Specification for radio interference measuring apparatus (CISPR 16). Last amended 1985.
BS 800: 1988 (EN 55 014)
Specification for radio interference limits and measurements for household appliances, portable tools and other electrical equipment causing similar types of interference (CISPR 14). Last amended 1990.
Will be withdrawn 31 Dec 1995, replaced by BS EN 55014: 1993.
See Chapter 6.
BS 833: 1970 (1985)
Radio interference limits and measurements for the electrical ignition systems of internal combustion engines (CISPR 12, CISPR 21).
BS 905 (EN 55 013 and EN 55 020)
Sound and television broadcast receivers and associated equipment: electromagnetic compatibility
Part 1: 1991 (EN 55 013: 1990): Specification for limits of radio interference (CISPR 13)
Part 2: 1991 (EN 55 020: 1988): Specification for limits of immunity (CISPR 20)
BS 1597: 1985
Specification for limits and methods of measurement of electromagnetic interference generated by marine equipment and installations.
BS 2316
Radio-frequency cables
Parts 1 and 2: 1968 (1981)
General requirements and tests. British Government Services requirements.
Part 3: 1969 (1981)
Cable data sheets (metric and imperial units).
BS 4727
Glossary of electrotechnical, power, telecommunications, electronics, lighting and colour terms
Part 1: Group 09: 1976: Radio interference technology (IEC 50: Chapter 902).
BS 4809: 1972 (1981) *Now obsolete replaced by* ***BS EN 55 011: 1991***
Radio interference limits and measurements for radio frequency heating equipment (CISPR 11).
BS 5049: 1987
Methods of measurement of radio interference characteristics of overhead power lines and high voltage equipment (CISPR18 - 2).
BS 5260: 1975 (1990)
Code of practice for radio interference suppression on marine installations.
BS 5394: 1988 (EN 55 015)

Specification for limits and methods of measurement of radio interference characteristics of fluorescent lamps and luminaires (CISPR 15). Last amendment 1990.
Will be withdrawn 31 Dec 1995, replaced by BS EN 55015: 1993.
BS 5406: 1988
Disturbances in supply systems caused by household appliances and similar electrical equipment.
Part 1: 1988 (EN 60 555-1)
Glossary of terms
Part 2: 1988 (EN 60 555-2)
Specification of harmonics (Chapter 6)
Part 3: 1988 (EN 60 555-3)
Specification of voltage fluctuations (Chapter 6).
BS 5602: 1978 (1990)
Code of practice for abatement of radio interference from overhead power lines (CISPR 1, CISPR 18-1, CISPR 18-3, CISPR Report 44).
BS 5783: 1987
Code of practice for handling electrostatic sensitive devices.
BS 6201
Fixed capacitors for use in electronic equipment
Part 3: 1982: Specification for fixed capacitors for radio interference suppression. Selection of methods of test and general requirements (IEC 384-14).
BS 6299: 1982
Methods of measurement of the suppression characteristics of passive radio interference filters and suppression components (CISPR 17).
BS 6345: 1983
Method of measurement of radio interference terminal voltage of lighting equipment. (Based on CISPR 15)
BS 6527: 1988 (EN 55 022: 1987)
Specification for limits and methods of measurement of radio interference characteristics of information technology equipment (CISPR 22). Several amendments pending.
Described in detail in Chapter 6.
BS 6651: 1990
Code of practice for protection of structures against lightning.
BS 6656: 1986
Guide for prevention of inadvertent ignition of flammable atmospheres by radio frequency radiation.
BS 6657: 1986
Guide for prevention of inadvertent initiation of electro-explosive devices by radio frequency radiation.
BS 6667: 1985
EMC for industrial-process measurement and control equipment (IEC 801):
Part 1: General introduction (implements CENELEC HD 481.1, IEC 801-1);
Part 2: Method of evaluating susceptibility to electrostatic discharge (implements CENELEC HD 481.2, IEC 801-2) will be withdrawn 31 Dec 1995;
Part 3: Method of evaluating susceptibility to radiated electromagnetic energy (implements CENELEC HD 481.3, IEC 801-3).
Described in detail in Chapter 7.
BS 6839: 1987
Mains signalling equipment:
Part 1: Specification for communication and interference limits and measurements. Last amended 1987. (Will be amended in line with EN 50 065-1: 1990, or replaced as BS EN 50 065-1.)
3G 100
Specification for general requirements for equipment for use in aircraft:
Part 4: Electrical equipment:
Section 2: 1980: Electromagnetic interference at radio and audio frequencies. Last amended 1985.
BS EN 50065-1: 1992
Signalling on low voltage installations.

BS EN 50 081-1: 1992
Electromagnetic compatibility - generic emission standard
Part 1: 1991: Residential, commercial and light industry.
Described in Chapter 6.
BS EN 50 081-2: 1994
Electromagnetic compatibility - generic emission standard
Part 2: 1994: Industrial.
Described in Chapter 6.
BS EN 50 082-1: 1992
Electromagnetic compatibility - generic immunity standard
Part 1: 1991: Residential, commercial and light industry.
Described in detail in Chapter 7.
BS EN 55 011: 1991
Limits and methods of measurement of radio disturbance characteristics of industrial, scientific and medical (ISM) radio-frequency equipment (CISPR 11).
BS EN 55 014: 1993
Limits and methods of measurement of radio disturbance characteristics of electrical motor operated and thermal appliances for household and similar purposes, electric tools and electric apparatus (CISPR 14).
BS EN 55 015: 1993
Limits and methods of measurement of radio disturbance characteristics of electrical lighting and similar equipment (CISPR 15).
BS EN 60601-1-2: 1993
Medical Electrical Equipment Part 1: General requirements for Safety- Collateral Standard, Electromagnetic Compatibility requirements and test (IEC 601-1-2: 1993).
BS EN 60801-2: 1993
Electromagnetic Compatibility for industrial process measurement and control equipment, Part 2: Electrostatic Discharge requirements (IEC 801-2: 1991). Supersedes HD 481.2/IEC 801-2: 1984.
BS EN 61000-4-7: 1993
Electromagnetic Compatibility (EMC) Part 4: Testing and measurement techniques. Section 7: General guide on Harmonics, Interharmonics Measurements and Instrumentation, for power supply systems and equipment connected thereto (IEC 1000-4-7: 1991)
BS EN 61000-4-8: 1994
Electromagnetic Compatibility (EMC) Part 4: Testing and measurement techniques. Section 8: Power frequency magnetic field immunity test; Basic EMC publication (IEC 1000-4-8: 1993)
BS EN 61000-4-9:1994
Electromagnetic Compatibility (EMC) Part 4: Testing and measurement techniques. Section 9: Pulse magnetic field immunity test; Basic EMC publication (IEC 1000-4-9: 1993)
BS EN 61000-4-10: 1994
Electromagnetic Compatibility (EMC) Part 4: Testing and measurement techniques. Section 10: Damped oscillatory field immunity test. (IEC 1000-4-10: 1993)
BS PD 6582
Guide to Generic EMC Standards
Drafts for development:
DD 158: 1987
Filters for mains signalling systems.
DD ENV 50140
Electromagnetic Compatibility. Basic immunity standard. Radiated radio frequency electromagnetic field. Immunity Test.
DD ENV 50141
Electromagnetic Compatibility. Basic immunity standard. Conducted disturbances induced by radio frequency fields. Immunity Test.
DD ENV 50142
Electromagnetic Compatibility. Basic immunity standard. Surge Immunity Test.
DD ENV 61000-2-2

Electromagnetic Compatibility (EMC) Part 2: Environment Section 2- Compatibility levels for low-frequency conducted disturbances and signalling in public low-voltage power supply systems (IEC 1000-2-2).

Appendix C

Listing of Principal DIN/DIN VDE EMC Specifications

DIN VDE 0228
Interference of telecommunication installations by electric power installations:
Part 1: 1987: General
Part 2: 1987: Interference by three phase installations
Part 3: 1988: Interference by ac traction systems
Part 4: 1988: Interference by dc traction systems
Part 5: 1987: Interference by high voltage dc transmission systems
E. Part 6:1992: Interference on information technology equipment; electrical and magnetic fields in the frequency range 0 to 10 kHz (E prefix, draft).
DIN VDE 0565
Specification for RFI suppression devices:
Part 1: 1979: RFI suppression capacitors
Part 2: 1978: RFI suppression chokes
Part 3: 1981: RFI interference filters up to 16A
Part 4: RFI suppression capacitors comprising a ceramic dielectric.
DIN VDE 0808-1: 1991
Signalling on low-voltage electrical installations; German version prEN 50 065 Part 1: 1989. Replaced by DIN EN 50065-1.
DIN VDE 0838
Disturbances in Supply Systems caused by household appliances and similar electrical equipment:
Part 1: 1992: Definitions, German version of EN 60555-1: 1987
Part 2: 1987: Harmonics, EN 60555-2, replaced by E.DIN EN 61000-3-2: 1994
Part 3: 1987: Voltage fluctuations, German version of EN 60555-3:1 987
See chapter 6 for EN 60555-2 and -3.
DIN VDE 0839-1: 1986
EMC; compatibility level of voltage in ac-systemsrated at up to 1000V.
DIN VDE 0839-10: 1989
EMC; judging immunity to conducted disturbances.
DIN VDE 0843
Electromagnetic Compatibility for industrial-process measurement and control equipment. German version of IEC 801 and HD 481:
Part 1: 1987 (IEC 801-1, HD 481.1): General
Part 2: 1987 (IEC 801-2: 1984, HD 481.2): Electrostatic discharge requirements, replaced by DIN EN 60801-2
E. Part 2: 1991 Electrostatic discharge requirements, intended to become equivalent to IEC 801-2: 1991, replaced by DIN EN 60801-2.
Part 3: 1988 (IEC 801-3: 1984, HD 481.3): Radiated electromagnetic field requirements.
E. Part 4: 1987 (IEC 65(CO)39: 1985): Fast transient requirements, will become equivalent to IEC 801-4: 1988.
DIN VDE 0846
EMC measurement apparatus:
Part 1: 1985: Harmonics of mains voltages and currents up to 2500 Hz.
Part 2: 1989: Flickermeter functional and design specifications; German version of IEC 868: 1986, replaced by DIN EN 60868: 1994, see Chapter 6.
E. Part 11: 1990: Signal generators.

E. Part 12: 1990: Coupling equipment.
E. Part 13: 1990: Measuring ancilliaries.
E. Part 14: 1990: Power amplifiers.
DIN VDE 0847
EMC measurement procedures:
Part 1: 1981: Conducted interference.
E. Part 2: 1987: Conducted immunity.
E. Part 4: 1987: Radiated immunity.
DIN VDE 0870-1: 1984
EMI; definitions
DIN VDE 0871: 1978
Limits of radio interference from RF apparatus and installations
Covers RF equipment for industrial, scientific and medical (ISM) and similar purposes and data processing equipment and electronic office machines. Limits of interference and measurement methods. Replaced by DIN VDE 0875-11: 1992. Remains valid to 31 Dec 1995. *See* Chapter 12 for details.
DIN VDE 0872-13: 1991
Radio interference suppression for sound and television broadcast receivers; limits and methods of measurement for radio disturbance characteristics of broadcast receivers and associated equipment; German version of EN 55 013: 1990
DIN VDE 0872-20: 1989
Immunity from radio interference of broadcast receivers and associated equipment; German version of EN 55 020: 1988
DIN VDE 0873
Radio interference characteristics of overhead power lines and high voltage equipment. 'Beiblatt' 1, 2 and 3 are identical to CISPR 18 parts 1, 2 and 3. Also:
Beiblatt 1: 1982: Measures against radio interference from electrical utility plants and electric traction systems; radio interference from systems of 10 kV and above.
Beiblatt 2: 1983: Measures against radio interference from electrical utility plants and electric traction systems; radio interference from systems below 10 kV and from electrical trains.
Beiblatt 3: 1991: code of practice for minimising the generation of radio noise
Part 1: 1982: measures against radio interference from electric utility plants and electric traction systems, 10kV and above.
DIN VDE 0875-1:1988
Radio interference suppression of electrical appliances and systems; limits and methods of measurement of radio interference characteristics of household electrical appliances, portable tools and similar electrical apparatus; (CISPR 14) German version of EN 55 014: 1987
DIN VDE 0875-2: 1988
Radio interference suppression of electrical appliances and systems; limits and methods of measurement of radio interference characteristics of fluorescent lamps and luminaires; (CISPR 15) German version of EN 55 015: 1987
DIN VDE 0875-3: 1988
Radio interference suppression of electrical appliances and systems.
DIN VDE 0875-11: 1992
Limits and methods of measurement of radio disturbance characteristics of industrial scientific and medical (ISM) RF equipment, EN 55011.
DIN VDE 0876
Radio interference measuring apparatus.
Part 1: 1978: Radio interference receiver with weighted indication and accessories.
Part 2: 1984: Disturbance analyser for the automatic assessment of interference produced by switching operations.
Part 3: 1987: Current probes for measuring radio interference.
DIN VDE 0877
Measurement of radio interference (CISPR 16):
Part 1: 1989: Measurement of radio interference voltages.
Part 2: 1985: Measurement of radio interference field strengths

Part 3:1980: Measurement of radio interference power on leads
Part 100: Specification for CISPR radio interference measuring apparatus for the frequency range 0.15 to 30MHz.
Part 101: Methods of measuring decoupling factors
DIN VDE 0878
EMC standard for telecommunications and information technology equipment:
Part 1: 1986: Radio interference suppression of telecommunication systems and apparatus; general specification
Part 2: 1988: Radio interference suppression of telecommunication equipment; equipment in telecommunication operating rooms.
Part 3: 1989: Limits and methods of measurement of radio interference characteristics of information technology equipment (CISPR 22), German version of EN 55 022 (*see* Chapters 6 and 11).
E. Part 30: 1989: National supplement to DIN VDE 0878 Part 3.
Part 200: 1992: Radio interference suppression of telecommunication systems and apparatus; method of measurement and limits.
DIN VDE 0879
Radio interference suppression for motor vehicles, vehicle equipment and internal combustion engines
Part 1: 1979: long distance interference suppression(CISPR 12).
Part 3: 1981: interference suppression for onboard radio reception.
Part 4: 1993: Methods of measurement to determinethe attenuation characteristics of radio interference suppressors of HV Ignition systems.

German equivalents to ENs, ENVs, are denoted by DIN EN, DIN ENV and are equivalent to BS ENs, DD ENVs *see* Appendix B.

Note: the 'E.' prefix denotes a draft standard, the normal form is, for example, E. DIN VDE 0878-30 : 1989.
For a complete list of German standards or the specific status of a particular standard, it is recommended that the reader should contact Technical Help to Exporters, British Standards Institution, Milton Keynes.

Appendix D

Principal US Commercial EMC Standards

Federal Communications Commission (FCC) documents

Code of Federal Regulations No. 47 Vol. 1, Office of the Federal Register National Archives and Records Administration, 1993:
Part 15 - Radio Frequency Devices, Described in detail in Chapter 12.
Part 18 - industrial, scientific and medical equipment (ISM), Regulation specifying electromagnetic emission limits for ISM equipment (CISPR 11). Described in detail in Chapter 12.
FCC/OST MP-4,
Methods of measurement of radio noise emissions from digital devices.
OST Bulletin No. 55
Characteristics of open field test sites.
OST Bulletin No. 62,
Understanding the FCC regulations concerning digital devices.

American National Standards Institute (ANSI)

ANSI/IEEE C37.90 - 1978

Relays and relay systems associated with electric power apparatus, section 9.3 'Surge withstand capability (SWC) waveshape and characteristics'.

ANSI C63:

Complete 1989 edition Electromagnetic Compatibility (published by the IEEE).

ANSI C63.2-1987

Standard for Instrumentatation - Electromagnetic Noise and Field Strength, 10kHz to 40GHz - specifications

ANSI C63.4-1988

Standard for Electromagnetic Compatibility - radio noise emissions from Low-voltage electrical and electronic equipment in the range of 10kHz to 1 GHz - methods of measurement.

ANSI C63.5-1988

Standard for Electromagnetic Compatibility - radiated emission measurements in electromagnetic interference (EMI) control - calibration of antennas.

ANSI C63.6-1988

Standard for Electromagnetic Compatibility - Open Area Test Site measurements - guide for the computation of errors.

ANSI C63.7-1988

Guide for construction of Open Area Test Sites for performing radiated emission measurements.

ANSI C63.12-1987

Standard for Electromagnetic Compatibility limits - recommended practice.

ANSI C63.4-1992

Methods of Measurement of Radio-Noise Emissions from Low Voltage Electrical and Electronic Equipment in the Range 9 kHz to 40 GHz.

ANSI C63.13-1991

Guide on the application and evaluationn of EMI power-line filters for commercial use.

Institute of Electrical and Electronics Engineers (IEEE)

IEEE Std 139-1988

IEEE Recommended practice for measurement of radio frequency emission from industrial, scientific and medical (ISM) equipment installed on user's premises (ANSI)

IEEE Std 140-1990 (Revision of IEEE Std 140-1950)

IEEE Recommended practice for minimization of interference from radio-frequency heating equipment (September 1990).

IEEE Std 473-1985

IEEE Recommended practice for an electromagnetic site survey (10 kHz to 10 GHz) (ANSI).

Scientific Apparatus Makers Association

SAMA standard PMC 33.1 - 1978

'Electromagnetic Susceptibility of Process Control Instrumentation'.

Note: many of the ANSI/IEEE standards are dual publications.

Appendix E

EMC Education

The W S Atkins report[2] highlighted the need for EMC education and training. It makes it clear that whilst the mismatch between test facilities and the numbers of products requiring testing can be addressed by investment, the available personnel to run such facilities is limited. Atkins[2] considers this to be 'the key limiting factor on expansion of capacity'. There has been no government initiated strategic response despite Atkins' warnings. The thinking behind the extension of the transitional period may also have included the idea that this period would allow adequately educated personnel to be generated gradually and obviate the need for a step change in numbers.

1 Tertiary education

Whilst there has been no strategic response from the Department of Education and Science (DES), tertiary education has responded to the need for EMC engineers by the introduction of EMC teaching into some undergraduate courses. This is known to be the case at the Universities of York, Hull Liverpool, Hertfordshire and Paisley. In addition to undergraduate education, Bradford and Huddersfield offer EMC as options on MSc courses and York and Hull ran a dedicated EMC MSc course.

The York/Hull Master's course commenced in 1990 and was the only course of its type not only in Europe but in the world. The philosophy and content of this course was described in a paper at the IEE 7th Conference on EMC [Riley *et al*, 1990[4]]. The course was available either for full time attendance for one year or part time for two years. The part time students received video tapes of lectures presented to the full time students and attended at either York or Hull for core weeks. Funding is being sought to set up a European Master's course based on the York/Hull formula. The final arrangements have yet to be agreed but it is proposed that the fundamentals will be taught in each of the cooperating institutions in the native language and the more specialised topics in English. The York/Hull masters course will be superseded by an Integrated Graduate Development Scheme (IGDS) 'A part time MSc in EMC and RF communications' sponsored by the Higher Education Funding Council (HEFC), this course will be available from October 1995.

2 EMC Continuing education

Whilst teaching initiatives will bear fruit in the future, in-service training (INSET) and Continuing Education (CE), for practising engineers is in great demand. Courses have been offered commercially by several organisations, including:

1)	R+B Enterprises	)US
	Bernard Keiser	)based
2)	I²T in conjunction with RFI	)
	Elmac Services in conjunction with Chase EMC	)all
	Universities of Paisley and Strathclyde	)UK
	University of Hertfordshire	)based
	ERA Technology	)
	The Institution of Electrical and Electronics	)
	Incorporated Engineers (IEEIE)	)
	York Electronics Centre (YEC)	)

The YEC has been offering short courses in EMC for over ten years. These courses have been very successful and are now run three times per year. Numbers on courses are limited to 20 people, as experience has shown this to be the optimum number of participants from an educational viewpoint to allow interaction with the lecturers.

One way in which larger numbers of engineers can receive in-service EMC education or training is by using Distance Learning (DL) techniques. These have the advantage that engineers are not required to be away from the workplace, also that a 'package' may actually be similar in cost to a single place on a short course and may be used by any number of engineers. Hatfield Polytechnic (University of Hertfordshire) produced an EMC DL (Distance Learning) package which based on videotape. This was produced under the DTI's Video Courses for Industry programme administered by Aston University.

The YEC has also produced an EMC video based DL package for the IEE [R04, 1992, 1995[5]]. This latter course is based on the well proven short course material and extensive use has been made of computer generated graphics and animated sequences. This course became available from late 1992 and was updated and further material added in February 1995.

The IEE Distance Learning course is modularised such that the material may be used in a variety of ways. For example the first two modules can be used as the basis for an awareness presentation. With alternative written support material the course may be used as the basis of a module on say a four module Master's course or for training technician engineers. It is also proposed to produce versions dubbed in other European languages.

3 Summary

Whilst there has been no strategic response from either the DTI or the DES, to the EMC education of engineers, higher education, the engineering institutions and industry have responded to the challenge of the perceived shortage of EMC skilled personnel. Provision has been made for both the short and longer term requirements:

i) Continuing education is being made available through a number of providers to retrain and educate practising electrical/electronics engineers, by means of short courses and video based distance learning materials.
ii) Postgraduate EMC courses now exist to supply 'EMC experts'.
iii) EMC is becoming a part of undergraduate and technician education.

Appendix F

Developments in Measurement Techniques

Prior to May 1989 when the EC Directive on EMC was officially adopted, few commercial EMC standards were legally binding, therefore technical errors within the standards themselves were not significant. The introduction of the EMC Directive and the legal requirement to comply with it has changed this. However, because of the requirement to implement standards quickly to enable manufacturers to self-certify, existing standards have had to form the basis of new relevant standards. Therefore as demonstrated in Chapters 5, 6 and 7 there have been large numbers of draft standards introduced and where deficiencies have been identified, these have been followed by an equally large number of amendments to these and to already existing standards.

From the commercial viewpoint manufacturers wish to minimise the testing that has to be carried out and, equally, they wish to minimise the actual cost of testing. This latter desire has led to the adoption of a number of new developments before they have been fully verified, *viz* the use of the 'Van Veen loop[1]', which has been adopted by amendment to CISPR 16[2]. Also the Garn[3] method has attracted attention as a method of 'calibrating' semi-anechoic chambers such that emission measurements can be corrected to those expected from an open field test site.

In earlier chapters the following have been identified as causes for concern:

i) the method known as site attenuation as a means of verifying the quality of an Open Field Test Site (OFTS)
ii) the resonances existing within screened enclosures
iii) the requirement for establishing a uniform EM field to a tolerance of 0 to +6 dB as defined in the proposed revision to IEC 801-3[4]
iv) the requirement generally for lower cost EMC test equipment

Developments addressing these concerns will be briefly described. In addition, the new developments of the 'GTEM' cell which is proving to be an alternative to the OFTS and screened room and the 'EMSCAN', which may be particularly effective as a production printed circuit board assessment tool, will also be reviewed.

1 OFTS quality

As discussed in Chapter 8 there are concerns about using the site attenuation method for measuring the quality of an OFTS. The main concern is that for this method tuned dipoles are set up at the usual measuring distance (3 m or 10 m) and that because the coupling between the antennas is very efficient the imperfections in the OFTS may have little effect on the measurements which are made.

Also measurements are made at discrete frequencies and full information about the site is not available for the intervals between these frequencies.

An instrument called the Comparison Noise Emitter (CNE) was developed at the University of York in 1988 for characterising screened rooms [Marshman and Marvin, 1989[11]]. This instrument was developed to production status by the YEC and some 40 have now been sold largely to test houses. This instrument is a self-contained broadband noise source generating an output over the frequency range x kHz to 1 GHz. The output when terminated in a 50 ohm load is shown in Figure 8.1. The unit is powered by an internal rechargeable battery and it can be used as a radiating source when fitted with a top-loaded monopole antenna. The primary use of the CNE is to ensure that measuring instruments are performing correctly and to check test set-ups on regular basis.

The CNE can also be used to compare the quality of open field test sites. Figure 8.2 showed the results from measuring the YEC OFTS. The output from the CNE whilst not 'flat', Figure 8.1, is continuous and 'smooth'. Examination of Figure 8.2 shows both features of the test site and the measurement set-up. Discontinuities at the frequencies 280, 300 and 650 MHz can be accounted for by features of the measuring antennas and the changeover from the Biconical to Log periodic antenna. Because the CNE generates continuous noise over the frequency range for which the OFTS is to be used and it can be used with the measuring instruments normally used on the site, imperfections in either the site or the measurement system can be identified. Thus the CNE offers a method for assessing the comparative quality of OFTSs and may be used in addition to the accepted site attenuation method.

NAMAS have carried out a survey of accredited OFTSs using a CNE. The results show a variation of +/- 4dB between the sites measured. It is likely that the CNE will continue to be used as a 'comparative standard' by NAMAS and may form part of the accreditation procedure for OFTSs in the future.

2 Screened room resonances

Chapter 8 discussed the problems associated with making measurements in screened enclosures. In summary, reflections from the boundaries of the enclosure and the cavity resonances which exist and which are dependent on the size of the enclosure, can produce variation in emission measurements of +/- 40 dB. This variation can be reduced by applying absorber to the walls and ceiling of the enclosure to reduce reflections and, as has been described the resonances can be damped by strategically positioning absorber or ferrite tiles. This latter method was devised at the University of York [Dawson and Marvin[12]] and is now commercially available.

Grace, Emerson and Cumming, offer ferrite tiles bonded to steel mounting plates, a design facility to calculate the number of tiles required and the tile pattern. As yet there are insufficient installations of different sizes to validate the technique fully in practice.

Characterisation of screened rooms and semi-anechoic chambers, particularly to improve repeatability of test methods and results, can be facilitated by use of the CNE described above. A further device is now available resulting from research at York, the spherical dipole swept emitter, SDSE. This device is a 100 mm diameter spherical dipole which receives a signal from a modulated fibre optic link. The light is received and demodulated within the sphere and the RF signal amplified and transmitted from the hemispheres which are separated by an insulating disc. The prototypes and first production units operated up to 250 MHz, the upper frequency being limited by the fibre optic link. A 1 GHz version has been developed and a version containing the broadband emitter from the CNE. The main advantage offered by the SDSE is that it is a calculable source [Marvin & Simpson[13]]. Unlike the CNE which can only be used for comparison measurements the SDSE has the potential for use as a 'transfer standard'.

3 Field uniformity for radiated immunity measurements

IEC 801-3 (revised)[10] calls for tolerances on field uniformity of 0 to +6 dB. These tolerances appear to have been chosen by CENELEC and the IEC on the basis that they represent the uniformity that is attainable. In practice it is very difficult to achieve these tolerances and measurement repeatability between test houses is of major concern.

It is therefore necessary from a practical viewpoint to establish the conditions required for the specified uniformity and to devise procedures which will ensure repeatability of measurements. It is also necessary from a fundamental viewpoint to establish the uniformity parameters required for meaningful immunity testing. Research was carried out at York into this fundamental approach, sponsored by the DTI. The former approach was the subject of work carried out by Radio Frequency Investigation also for the DTI. The results of this work were made public in July 1992.

4 Low cost EMC test equipment

The Atkins[2] report highlighted the shortfall in test facilities to meet the compliance demands of the

EMC Directive, *see* Chapter 9. One response is the establishment of more EMC test houses, the alternative is the establishment of in-house facilities to provide at least pre-compliance confidence testing and product development testing. Whilst the do-it-yourself approach is generally taken by the UK engineering industry in other areas of technology, the cost of EMC test equipment and facilities tends to be prohibitive except for larger companies, leaving SMEs with no other choice than using a third party test house.

The costs of OFTS and screened room/anechoic chamber facilities were discussed in Chapter 8. The costs indicated range from £10 k for a basic OFTS to a facility costing of the order £1 million+. If emission testing only is considered then a suitable spectrum analyser alone is likely to cost £10 k. In addition to this the user will need antennas and a LISN. For immunity measurements, other than for frequencies below 200 MHz where a TEM cell can be used, a screened enclosure is essential if disturbance to other equipment users is to be prevented. A typical screened room when erected is likely to cost a minimum of £20 k, in addition valuable laboratory or factory space will have to be made available.

A number of companies now offer low cost emission measurement equipment for pre-compliance or development testing. In addition these instruments may be useful for production line comparative testing. Low cost immunity test equipment is also now emerging having a limited capability. A number of manufacturers are offering lower cost alternatives to the conventional screened room, these include conductive fabrics to line an existing room within a building or to make a screened tent.

The author and the YEC are involved with two solutions to the problems identified:

i) the 'EASY 1' (Emission Assessment System York number 1) a system built around Farnell Instruments spectrum analyser type SSA1000 and

ii) the 'RFI Portastor', a relocatable screened room, a joint development with Portasilo Ltd, York, part of the Shepherd building group.

The 'EASY 1' — emission assessment system

The EASY 1 is a complete emissions assessment system suitable for pre-compliance or development work. It consists of the Farnell SSA1000 spectrum analyser controlled *via* an IBM compatible personal computer using menu driven software, a novel 30 MHz to 1 GHz broadband antenna and a tripod capable of fixing the height of the antenna between 1 m and 2 m. This system can be used for evaluating radiated emissions on a makeshift site. Conducted emissions are measured using a LISN and a 30 MHz to 1 GHz near field probe has been designed for diagnostic work. A manual is provided giving guidance on choice of makeshift sites, standards and common EMC problems.

The principal advantages offered by the EASY [Wainwright[8]] are:

i) the single antenna, offering a performance comparable to biconical and log periodic antennas when used on a 3 m site, but obviating the need to change antennas during emission measurements, *see* Figures A.1 and A.2.

ii) the software has been designed to set-up all the spectrum analyser settings appropriate to the common emission standards, which are stored as data sets and uses the Windows™ 3.1 operating environment. The use of menus means that electronics engineers and technicians unfamiliar with EMC measurements or performing such measurements infrequently, are able to make sensible measurements as the software provides the prompts and sets up the appropriate parameters on the analyser such as span and resolution bandwidth.

iii) the cost of the system is about £12 k and therefore should be within the reach of many SMEs. Also as a low cost system it may find applications with larger companies for production test equipment and in development areas.

iv) unlike many of the low cost emission systems the EASY 1 offers the full capability of diagnostic, conducted and radiated emission testing and the spectrum analyser uses the CISPR 16 bandwidths of 9 kHz and 120 kHz giving the system a low noise floor.

The 'RFI Portastor'

The 'RFI Portastor' is based on Portasilo's 'Portastor', a relocatable building for secure storage. This structure is of all steel welded construction and is completely weatherproof. To develop the Portastor into a screened enclosure has involved the addition of a gasketted door, steel floor and increasing the frequency of the spot welding to reduce the gaps. The objective was to produce a relocatable screened room having sufficient attenuation to allow radiated immunity measurements at 10 V/m to be performed, rather than the high levels of attenuation offered by the established screened room manufacturers for rooms which were originally designed for the military EMC market.

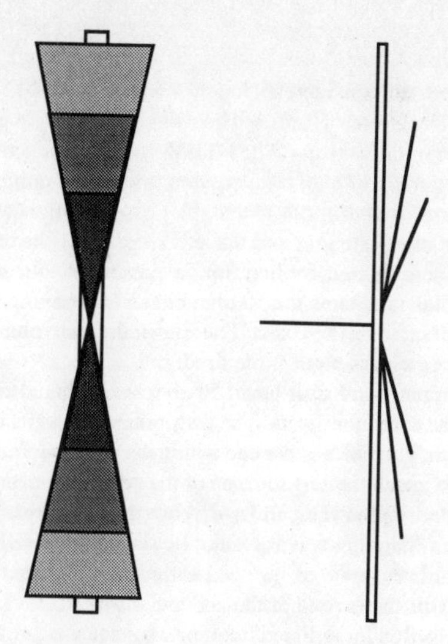

Figure A.1 EASY 1 Antenna

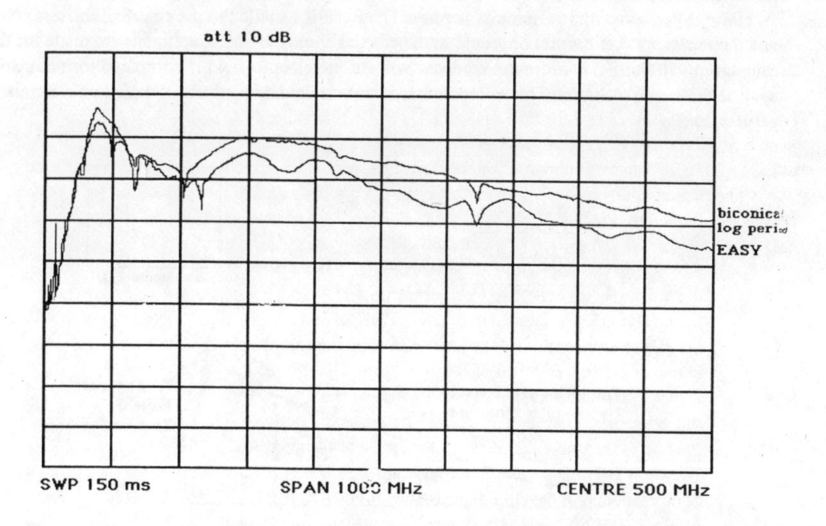

Figure A. 2 Comparison of EASY antenna and biconical/log-periodic

It was assumed that at this field strength the screening should be adequate to limit emissions from the enclosure to the Class A emission limit defined by EN 55 022. To minimise resonance effects, Ferrite damping tiles are provided [Dawson and Marvin[6]]. This enclosure is delivered from the factory completely tested and can be erected on any suitable ground within a factory environs, being fully weather proof, thereby relieving the need to find space within the factory. It is also lower in cost than comparable enclosures. An attractive feature is the provision of a vestibule enclosing the RFI door, this protects the door from the weather it also provides a working space for the test engineers and the instrumentation.

5 The GTEM™ cell

A device which has created considerable interest is the 'GTEM' cell or 'Gigahertz Transverse ElectroMagnetic' cell. The interest is aroused because the GTEM in principle offers an environment for both emission and immunity testing. The GTEM is a development of the TEM cell and as shown in Figure A.3 it is a section of 50 ohm transmission line with a unique geometry. A 50 ohm coaxial transmission line is transformed into a rectangular cross-section with a ratio of 2:3 in height to width. The cell is flared along the longitudinal axis to increase the cross-sectional dimensions. The septum, or centre conductor is transformed from a circular conductor to a flat plate, located above the centre of the cell. This maintains the 50 ohm characteristic impedance, but allows the test volume underneath the septum to be increased. The size of the test volume is proportional to the length, hence for larger EUTs a greater length is required.

The septum is terminated in a distributed 50 ohm load, printed circuit boards are used mounting several thousand discrete carbon resistors. The distribution of the resistor values matches the current distribution in the septum. The fields generated within the cell are terminated in an RF absorber having characteristics chosen to match the performance of the cell as a function of frequency.

EM fields generated within the cell by driving the RF connector are terminated in the load boards and the absorber. Standing waves cannot be set-up because of the structure of the cell; EM waves propagated towards the apex are damped as the tapering structure acts as a cut off waveguide, waves propagated towards the far end of the cell are absorbed.

When the GTEM is used for radiated immunity tests the test volume is bounded by dimensions of the height either far or near as shown in Figure A.3. Within the test volume a field uniformity of +/- 1 dB is claimed [Osburn, 1990[9]]. The signal generator/RF amplifier combination is connected to the coaxial connector and adjusted to provide the required field strength.

For emission testing reciprocity applies. The EUT is installed in the centre of the test volume and the receiving equipment connected to the coaxial connector. Measurements are made for three orientations of the EUT. From the received voltage the electric field is computed for comparison against the required limits. This computation is not straightforward and is based on a number of assumptions.

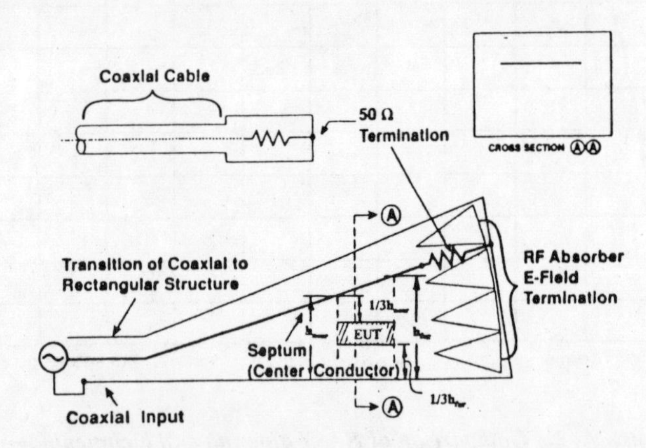

Figure A. 3 GTEM cell

A simplified description of the technique follows: at each measurement frequency the three orthogonal voltage measurements are summed to produce a resultant voltage. From this is derived equivalent electric and magnetic dipoles which are then mathematically placed over a perfect ground plane, the measurement distance is specified (*eg.*, 3 m or 10 m) and iterations are made of the vertically and horizontally polarised electric field strength over the scan height of 1 to 4 m. The maximum field strength is determined from the computations and compared with the specified limit. This is repeated over the whole frequency range. The time claimed for performing such a measurement is 30 minutes [Osburn,1990[9]].

The GTEM appears to offer significant advantages over the screened room and OFTS described in Chapter 8, these are respectively, reflections and standing waves are theoretically eliminated and the external ambient signals are excluded. Also the time to perform measurements is significantly reduced which is particularly attractive to manufacturers requiring to test a large number of sample products for production control.

The cost of the GTEM is certainly lower than for an anechoic chamber or sophisticated OFTS. However the technique has not been universally accepted. Limited acceptance has been given by the FCC to emission measurements made in a GTEM and when implemented, the revision to IEC 801-3 allows the use of any type of test facility as long as the field uniformity requirements are met.

The major drawback is the size of EUT which can be accommodated and it is likely that its use will be restricted to personal computers and domestic appliances.

6 EMSCAN™

The EMSCAN was developed by Bell-Northern Research (BNR) and is now marketed by the EMSCAN Corporation. It is essentially an array of loop antennas coupled to a spectrum analyser and controlling personal computer. A printed circuit board (pcb) is placed on the unit and exercised, at the same time a scan is made of the near field electromagnetic emissions and the sources of emission identified. These sources are then displayed rather like a contour map on the PC monitor.

The EMSCAN was originally developed as a development tool to enable BNR to minimise the emission sources from pcbs and therefore minimise the compliance testing required for finished products. However a major role now identified for the EMSCAN is production EMC testing of pcbs [Xavier and James, 1991[10]] The EM 'contour' map for a pcb used in a certified product, can be stored in the PC and then production pcbs can be measured, the emission sources identified and compared with the sample. Clearly tolerances will have to be specified.

The advantage of this sort of testing is that it can follow a functional test using a conventional 'bed of nails' to connect the pcb to the automatic test equipment (ATE) and may identify such problems as 'dry joints'. A dry joint may perform satisfactorily on a bed of nails and then subsequently cause a finished product to malfunction, but a dry joint will have a different RF performance to a good soldered joint which will show on the contour map. Therefore the EMCSAN technique is being developed with a view to replacing conventional ATEs; if a pcb is functioning correctly its emission map will be the same as for a known good board, if however it is not functioning correctly its map will be different, but the advantage of the EMSCAN is that physical electrical contact with the pcb is not required.

References

1. J R Bergervoet and H Van Veen 'A large loop antenna for magnetic field measurements' 8th International Symposium on EMC, Zurich, March 1989, pp 29-34
2. Draft - Proposed revision of CISPR Publication 16 'CISPR specification for radio interference measuring apparatus and measurement methods. Part 2 - Methods of Interference and immunity measurement. Clause 7X: Measurements in a Loop Antenna system', BSI draft 90/28294 DC, August 1990
3. H Garn, J Rasinger, W Mullner and E Bonek 'An improved method for the performance evaluation of radiated emission test sites by a comb generator reference', IEE 7th International Conference on EMC, Conference Publication Number 326, August 1990
4. 90/29283 DC 'Draft - revision of IEC publication 801-3: electromagnetic compatibility for industrial-process measurement and control equipment Part 3: immunity to radiated radio frequency electromagnetic fields', BSI, September 1990
5. Marshman and Marvin 'RFI Radiated Emission Regulations - A Measurement Technique to Assist Product Development', Enigma Variations Conference, May 1989
6. Dawson and Marvin 'Alternative Methods of Damping Resonances in a Screened Room', IEE Sixth International Conference on EMC, 1988
7. A C Marvin and G C Simpson 'Emission measurements in screened rooms: a preliminary calibration procedure for the frequency range 1to 30 MHz', EuroEMC'91, October 1991
8. N J Wainwright 'Effective low cost pre-compliance emission measurements', 16th Automated RF & Microwave Measurement Society Conference, Heriot-Watt University, March 1992
9. J Osburn 'New test cell offers both susceptibility and radiated emissions capabilities', RF Design, August 1990, pp 39-47
10. S Xavier and D James 'EMSCAN a tool for evaluating the EMC performance of printed circuit packs', EURO-EMC'91, London, 8-10 October 1991

Appendix G

CAE/CAD for EMC —a Review of Computational Electromagnetics Techniques [Porter 1992]

1 Introduction

Over the past twenty years, an increase in both the speed and capacity of computer systems has revolutionised most engineering disciplines. Electrical and electronics engineering is now dependent upon computer aided design (CAD) and computer aided engineering (CAE) systems. For example, these are used for circuit schematic capture and pcb layout, and also circuit simulation. In the area of electromagnetic engineering, computer tools have been developed specifically for antenna and microwave design. Overall there is now a movement towards rigorous and integrated engineering methodologies. EMC is a relatively new area of application for CAE techniques. However, the importance of this discipline has encouraged research and development over the last few years, resulting in a wide variety of tools.

2 The role of CAE in EMC design

Traditionally, EMC has been considered at the post-design stage. Having already been built, it is only then that equipment is tested to see whether or not it conforms to the relevant standards.

This can prove very expensive in terms of time, costs, and the potential need for retrofit modifications. Simulating a piece of equipment is potentially much faster and cheaper than taking a prototype or existing piece of equipment to a test-house. More importantly, it allows the engineer to 'look into' the equipment and see where currents and fields are largest; which is almost impossible with physical testing.

CAE tools can influence a design before it is finished or has been built, whether this is simulating an initial mechanical design, individual circuit boards, or the whole design including both mechanical and electrical aspects.

When simulating an initial mechanical design, white noise or single frequency sources may be used to excite the structure in place of circuit boards which may not yet exist. It is thus possible to vary the simulated design to ensure optimum EMC performance in the final design, for example, by varying the position of the sources to determine the best location of circuit boards. When simulating individual circuit boards, CAE tools can indicate whether components need to be moved and tracks re-routed to give a lower radiation profile. Simulating a whole design allows for the early identification of EMI problem areas with consequent early electrical and mechanical redesign. It allows complete assessment of all potential electromagnetic interactions, something which is impossible with conventional single-circuit-board and mechanical-design simulations or from the results of performing measurements.

3 Sources of EMI within a design

There is a vast range of elements that needs to be modelled including common-mode and differential-mode currents, and associated coupling paths, wires, surfaces, bodies, all necessary interconnections, nonlinear materials, anisotropic materials, enclosures, and open spaces. These present a formidable challenge for any one modelling technique.

4 Three fundamental modelling methodologies

Having defined the problems, the fundamental methodologies behind Computational Electromagnetic (CEM), methods will be reviewed; that is analytical, numerical and expert systems.

Most electromagnetic problems can be reduced to a form where any interactions are first described using a number of partial differential equations, and then applying appropriate boundary conditions *eg* the driving voltages.

4.1 Analytical methods

In order to solve a problem analytically the design must first be reduced to some simplified geometry to which a closed-form solution is applied. The advantage of this method lies in its simplicity. Provided that the important electromagnetic interactions can be anticipated, it is possible to ignore irrelevant aspects of a problem. Analytical methods can be very simple, and can run easily on small computers such as PC's.

Unfortunately, for many EMI problems the important interactions are often unpredictable and analytical methods are usually insufficient to give reasonable predictions.

4.2 Numerical methods

Numerical methods provide a means to solve the fundamental electromagnetic equations directly, by applying appropriate boundary conditions for the design geometry. There are many different techniques available to tackle a wide range of problems, however, the specific technique chosen for use in a particular CAE tool will often limit the actual range of problems the tool can be used for.

The fundamental advantage of numerical methods is that they do not make assumptions as to which electromagnetic interactions will be most significant in any particular design. Numerical methods analyse the whole of the design geometry without having to make major simplifications.

Because numerical methods are complex, they often require significantly more computation than analytical methods or expert systems. They also require a thorough knowledge of the design and thus either a large quantity of input data, or an interface with an existing CAE system. It is also difficult to find a single technique that is well suited to a full range of problems.

Numerical methods themselves fall into two categories, finite methods and integral methods. Maxwell's equations may be expressed in two forms. The differential form relates electromagnetic quantities at a point, to quantities in its vicinity, through a set of interactions. The integral form describes quantities throughout all space. Each of these formulations of Maxwell's equations has given rise to a broad area of research within CEM. The differential form is the basis of finite methods, while the integral form is the basis of integral methods. In both cases the design has to be initially sub-divided into smaller units, or discretised, to allow its representation within a computer. The various numerical methods are:

Finite methods

- finite element
- finite difference (both time and frequency domain)
- transmission line matrix
- uniform theory of diffraction

Integral methods

- moment (method of weighted residuals)
- generalised multipole.

Many of these techniques are useful for a wide range of applications, but none currently fulfils all of the necessary requirements. There has been considerable research into this and several combinations of techniques have been produced, but few software packages are currently available. An example is the York Electromagnetic Theory Implementation, which combines the transmission line modelling of printed circuit boards with moment method techniques for modelling attached structures, such as cabling and enclosures.

Other examples of packages using numerical methods are the Numerical Electromagnetics Code (developed by the Lawrence Livermore National Laboratory) and the General TLM Electromagnetics Code (developed by Auburn University).

4.3 Expert systems

Within an expert systems package there is no formal, rigorous calculation of the electromagnetic interactions. Instead, an estimate is made of the parameters of interest based on a rules database.

An advantage of such an approach is that it gives a very fast estimate of the parameters needed, although not as accurately as the methods mentioned earlier. Expert systems are suitable for insertion into automatic procedures for system design and board layout in order to allow some account to be taken of EMC considerations, an example is the BNR 'PC EM SIM' software which complements the CBDS 5 pcb layout package.

Expert systems are limited by the accuracy and appropriateness of their rather simplified rules database and model the complex interactions of EMI sources rather poorly.

5 Conclusions

No firm conclusions can be drawn as to which techniques will ultimately satisfy all of the requirements for the successful prediction of the EMC performance of electrical and electronic equipment. It is likely to be a hybrid technique combining more than one method. Computer aided engineering for EMC is becoming a reality and will be an indispensible tool for all electrical and electronics engineers within the foreseeable future [Porter, 1992: S J Porter 'CAD for EMC, Module 4, programme 4, Electromagnetic Compatibility (with particular emphasis on EC Directive 89/336/EEC)', IEE Distance Learning, 1992].

Appendix H

Technical Construction File Suggested Contents

Part I: Description of the apparatus:

i) Identification of apparatus;

(a) brand name;

(b) model number;

(c) name and address of manufacturer or agent;

(d) a description of the intended function of the apparatus;

(e) for installations - physical location;

(f) external photographs;

(g) any limitation on the intended operating environment.

ii) A technical description;

(a) a block diagram showing the interrelationship between the different functional areas of the apparatus;

(b) relevent technical drawings, including circuit diagrams, assembly diagrams, parts list, installation diagrams,

(c) description of intended interconnections with other products, devices etc;

(d) descriptions of product variants.

Part II: Procedures used to ensure conformity of the apparatus to the Protection requirements:

i) Technical rationale

A brief explanation of the rationale behind the inclusion of design aspects relating to the equipment, the test data provided and the balance between them.

ii) Detail of significant design aspects

(a) Design features adopted specifically to address EMC problems;

(b) relevant component specifications, eg the use of a cabling product known to minimise EMCproblems

(c) an explanation of the procedures used to control variants in the design together with an explanation of the procedures used to assess whether a particular change will require the apparatus to be retested;

(d) details and results of any theoretical modelling of performance aspects of apparatus.

iii) Test Data

(a) A list of the EMC tests performed on the product, and test reports relating to them, including details of test methods, etc;

(b) an overview of the logical processes used to decide whether the tests performed on the apparatus were adequate to ensure compliance with the protection requirements of the EMC Directive

(c) a list of tests performed on critical sub-assemblies and test reports or certificates relating to them.

Part III: Report or Certificate from a Competent Body

The report from the Competent Body may mirror the content of parts I and II of the TCF assembled by the manufacturer. hence it could:

(a) refer to the exact build state of the apparatus assessed, cross referencing with Part I;

(b) comment on the technical rationale;

(c) state the work done to verify the contents and authenticity of the design information in the TCF, cross referencing with Part II(ii);

(d) comment where appropriate on the procedures to control variants, and on environmental, installation and maintenance factors which may be relevant;

(e) contain an analysis of the tests performed either by the manufacturer, a third party, or the Competent Body itself, and the results obtained, so as to assess whether the tests indicatecompliance with the protection requirements of the EMC Directive, cross referencing with part II(iii).

TCF Parts I and II should be written by the manufacturer in cooperation or consultation with the Competent Body. The Competent Body report should therefore repeat little of the information contained in Parts I and II.

At the end of the report a detachable certificate will be supplied. This can be used by the manufacturer to indicate compliance when it is inappropriate to submit the full TCF. It is possible that Parts I and II 'speak for themselves' and the Competent Body may prepare a certificate only.

The user should note that the manufacturer is ultimately responsible for the declaration of conformity of products complying via the TCF route. The role of the Competent Body is to assert that the information contained within the TCF is consistent with conformity. It is the manufacturers reponsibility to ensure the information is correct and that subsequent production units are consistent with it.

Appendix I

Design practices used in large installations

Most EMC design practices are common sense to the professional electrical or electronics engineer and follow established good practice resulting from experience. Following guidelines during the design phase will minimise the likelihood of poor EMC, however verification by quantitative testing will be necessary to ensure system reliability.

Segregation

At system level placing of sensitive equipment adjacent to power conversion equipment should be avoided.

Within sub-systems use screened enclosures and separate control electronics from higher power apparatus, ensure high speed cicuits are located towards the centre of an enclosure with lower speed circuits adjacent to the enclosure faces.

Cabling may be the dominant coupling path for electromagnetic interference and it is necessary to categorise circuits according to their likely susceptibility or potential to be a major emitter.

Possible categories are:

Low voltage high speed digital logic connections
analogue voltage signals *eg* from transducers
low voltage control circuit connections *eg* relay logic
auxiliary power cabling
special cases *eg* supply cables

Having categorised the different circuit types the rule is never mix different categories in the same wiring harness or connector without additional screening between them. Where direct links are made between 'noisy' and sensitive circuits these must be filtered for both common and differential mode coupling. Ideally direct connections will be avoided by use of transformer or opto coupling techniques.

Cable routing

The electrical systems designer should assume that the mechanical layout designer will connect equipment by the shortest or most tidy route unless otherwise directed.

Explicit instructions should be given for:

- cables forming outward and return pairs which should be kept together, alternatively twisted pair cables should be specified
- circuits where there may be a single outward cable and several returns which may be safely bundled together
- linking three or more systems sharing common conductors
- ensuring that signal or control cables always cross power cabling at right angles to mini mise magnetic coupling

Connectors

Requirements for connectors:

- signals of different categories should not be mixed without the use of screening or adequate physical separation
- crosstalk between connector pins should be minimised however this is likely to be a function of the type of connector used
- where screened cables are terminated at connectors care must be taken to termninate the screen adequately, for example a full 360 degree termination of a coaxial sheath. Pigtail connections should be avoided, where there is no alternative the pigtail lenght should be as short as possible
- where more than one screened cable is terminated at a connector the screen isolation requirements must be specified.
- where the connector shell is used to continue a screen, a screw style connector should be used in preference to a bayonet type.

Earthing

There are two reasons for earthing or grounding electrical or electronic equipment:

1) the ''safety earth' provides a controlled route for hazardous fault currents and ensures that all exposed metallic or electrically conducting surfaces are at the same potential, so that a person will not experience an electric shock by touching two parts of the structure together
2) to provide a reference potential for the system and connections to it.

Care must be taken to avoid 'ground loops'. These are formed when the earthing arrangement produces continuous loops. Stray magnetic flux will link with the low impedance loop and currents are generated which may disrupt the operation of circuits connected to the earth system.

Earth loops may be avoided by using:

- 'Star' point earthing. where all the earth connections are brought back to asingle point
- by galvanically isolating circuits using isolation transformers, opto-isolators or fibre optics, however such techniques should not compromise safety earth requirements.

Common Impedance coupling

Where two circuits share a common earth return 'common impedance coupling' may result between the circuits, this may be minimised by:

- not allowing noisy and susceptible circuits to share common cable runs
- ensuring that a low impedance bond occurs at the nodes where earth cables finally meet
- using differential circuits which do not depend on local earth as a reference potential.

Ground planes

Electromagnetic interference can be minimised by locating cables or components close to an earthed structure. Wiring harnesses located close to earth will be less affected by major disturbances. A voltage surge propagating throughout the earth will affect signal lines and the earth structure together. Because they are closely coupled both will experience a similar large common mode disturbance, but the differential mode disturbance which is more likely to affect the operation of the circuits will be much reduced.

Screening

Screening or shielding is placing a physical barrier between circuits to reduce electromagnetic coupling. The effects of electric fields can be minimised by using a high conductivity barrier with a low impedance to ground, for example, placing circuits within a metal enclosure. Care must be taken to ensure that a good earth bond exists between the enclosure and earth otherwise noise diverted by the screen may couple to the victim circuit by common impedance coupling. Also holes and apertures in screens should be avoided to prevent electromagnetic leakage. For frequencies up to 1 GHz the maximum allowable hole size is 16 mm.

For low frequency magnetic fields typically less than 100 kHz, high permeability materials such as mumetal and steel will concentrate the field within the bulk of the material. However as the frequency rises non-magnetic materials such as aluminium become effective.

Screened cables are used to provide protection to low power signals or to attenuate emissions from noisy circuits. However consideration must be given to whether the screen is to be terminated at one or both ends.

If the screen is grounded at one end, it should be grounded at the signal source and will provide a screen to electric fields so long as the impedance of every part of the screen to earth remains low. There will however, be no protection to low frequency magnetic fields.

If the cable screen is grounded at both ends a ground loop or common impedance path must be avoided, but protection will be provided against low frequency magnetic fields. However large currents may flow in the screen.

A compromise for the designer may be to ground solidly at one end and connect the other end via a capacitor, typically 10 to 100 nanofarad, to ground. At low frequencies the screen appears to be grounded at one end preventing the flow of large currents in the screen, but at high frequencies appears to be grounded at both ends.

Suppression

Particularly relevant to large systems is the use of voltage clamping devices for suppressing relay and contactor coils. It should be noted that whilst it is desirable to suppress at source surges from these, their dynamic performance may be affected as the opening of a relay for example, will be slowed down.

Index

It is recommended that the reader uses the comprehensive contents list as a starting point, as each subject area must be considered in context with reference to the implications. The index should be used as a secondary source as it provides an index of definitions which may be referred to should a chapter be read in isolation. The index also references the pages where a subject is covered in detail but due to the legally binding nature of the EMC Directive it is important to research each area thoroughly within its context.

Chris Marshman is a recognised authority on the EMC Directive. A Member of the Institution of Electrical Engineers and a Chartered Engineer since 1980. Over the last 9 years he has established an EMC consultancy and measurement service, including test facilities, at the University of York and a programme of regular EMC short courses and workshops for industry and commerce, which is now combining with the Department of Electronics' EMC research group to form the EMC Technology Centre. He also lectures on the subject of the EMC Directive for the EMC MSc course.

He has made an extensive study of the EMC Directive and the Harmonised Standards and their implications for the University's industrial clients and is regularly invited to speak on this subject at conferences and to trade associations.

Chris was also the executive producer for the Institution of Electrical Engineers' EMC video-based distance learning package which has been recently updated.